Anthropocene Antarctica

Anthropocene Antarctica offers new ways of thinking about the 'Continent for Science and Peace' in a time of planetary environmental change. In the Anthropocene, Antarctica has become central to the Earth's future. Ice cores taken from its interior reveal the deep environmental history of the planet and warming ocean currents are ominously destabilising the glaciers around its edges, presaging sea-level rise in decades and centuries to come. At the same time, proliferating research stations and tourist numbers challenge stereotypes of the continent as the 'last wilderness.' The Anthropocene brings Antarctica nearer in thought, entangled with our everyday actions. If the Anthropocene signals the end of the idea of Nature as separate from humans, then the Antarctic, long considered the material embodiment of this idea, faces a radical reframing.

Understanding the southern polar region in the twenty-first century requires contributions across the disciplinary spectrum. This collection paves the way for researchers in the Environmental Humanities, Law and Social Sciences to engage critically with the Antarctic, fostering a community of scholars who can act with natural scientists to address the globally significant environmental issues that face this vitally important part of the planet.

Elizabeth Leane is Professor of English at the School of Humanities/Institute for Marine and Antarctic Studies, University of Tasmania.

Jeffrey McGee is Senior Lecturer in Climate Change, Marine and Antarctic Law at the Faculty of Law/Institute for Marine and Antarctic Studies, University of Tasmania.

Routledge Environmental Humanities

Series editors: Scott Slovic (University of Idaho, USA), Joni Adamson (Arizona State University, USA) and Yuki Masami (Kanazawa University, Japan)

The *Routledge Environmental Humanities* series is an original and inspiring venture recognising that today's world agricultural and water crises, ocean pollution and resource depletion, global warming from greenhouse gases, urban sprawl, overpopulation, food insecurity and environmental justice are all *crises of culture*.

The reality of understanding and finding adaptive solutions to our present and future environmental challenges has shifted the epicenter of environmental studies away from an exclusively scientific and technological framework to one that depends on the human-focused disciplines and ideas of the humanities and allied social sciences.

We thus welcome book proposals from all humanities and social sciences disciplines for an inclusive and interdisciplinary series. We favour manuscripts aimed at an international readership and written in a lively and accessible style. The readership comprises scholars and students from the humanities and social sciences and thoughtful readers concerned about the human dimensions of environmental change.

Anthropocene Antarctica

Perspectives from the Humanities, Law and Social Sciences

**Edited by
Elizabeth Leane and Jeffrey McGee**

LONDON AND NEW YORK

First published 2020
by Routledge
2 Park Square, Milton Park, Abingdon, Oxon OX14 4RN

and by Routledge
52 Vanderbilt Avenue, New York, NY 10017

Routledge is an imprint of the Taylor & Francis Group, an informa business

First issued in paperback 2021

British Library Cataloguing in Publication Data
A catalogue record for this book is available from the British Library

Library of Congress Cataloging-in-Publication Data
A catalog record has been requested for this book

ISBN: 978-1-138-36759-3 (hbk)
ISBN: 978-1-03-208915-7 (pbk)
ISBN: 978-0-429-42970-5 (ebk)

Typeset in Bembo
by Taylor & Francis Books

Contents

Illustrations

Acknowledgements

The editors wish to acknowledge the generous support for this project provided by the University of Tasmania, especially the College of Arts, Law and Education, the Institute for Marine and Antarctic Studies and the Marine, Antarctic and Maritime Research Theme; and the Humanities and Social Sciences Expert Group and History Expert Group of the Scientific Committee of Antarctic Research. All provided funding and/or in-kind support towards a 2017 conference which was the initial impetus for this collection. The editors also wish to acknowledge the generous support of the Australian Research Council, which has funded Professor Leane's Future Fellowship (FT120100402) over recent years. However, none of these groups have responsibility for the ideas expressed in the collection.

The editors express special thanks to our outstanding research assistant Dr Jacqueline Fox, whose excellent work and collegiality were invaluable during the later stages of preparation of this collection. Many thanks also to Julia Pollacco from Taylor and Francis for her support and understanding during writing and production of this volume, to Kristina Wischenkamper for her careful copy-editing, and to Edward Gibbons for steering us through the production stage.

Elizabeth Leane is grateful to Damian, Zac and Tessa for their unswerving support and encouragement while this collection was being put together. Jeffrey McGee wishes to thank his wife Maree and sons Connor, Lucas and Declan for their understanding of his family absences during the many weekends and late nights needed to produce this collection.

Contributors

Professor Sanjay Chaturvedi is Dean of the Faculty of Social Sciences, Department of International Relations, South Asian University, New Delhi. He specialises in theories and practices of Geopolitics and IR, with special reference to Polar Regions and the Indian Ocean Region. Professor Chaturvedi is also the Regional Editor of *The Polar Journal* (Routledge) and Member, International Executive Committee (ex officio) of the SCAR Antarctic Humanities and Social Sciences Expert Group (Geopolitics).

Dr Alan D. Hemmings is a self-employed Polar Specialist and an Adjunct Associate Professor at Gateway Antarctica Centre for Antarctic Studies and Research at the University of Canterbury in Christchurch, New Zealand. His research and publications focus particularly on the geopolitics and governance of the Antarctic and the 'Greater Southern Ocean'.

Adrian Howkins teaches environmental history at the University of Bristol in the United Kingdom. His research focuses on the history of the Polar Regions, and he is currently working on an environmental history of the McMurdo Dry Valleys, Antarctica. He is a co-PI on the McMurdo Dry Valleys Long Term Ecological Research site, and would like to acknowledge the support from the National Science Foundation (grants 1637708 and 1443475), which made possible the research for this chapter.

Elizabeth Leane is Professor of English at the University of Tasmania, where she holds an Australian Research Council Future Fellowship split between the Institute for Marine and Antarctic Studies and the School of Humanities. She is a chief officer of the Standing Committee on Humanities and Social Sciences (Scientific Committee on Antarctic Research) and is the Arts and Culture editor of *The Polar Journal*. Her publications include *South Pole: Nature and Culture* (Reaktion Books, 2016) and *Antarctica in Fiction* (Cambridge University Press, 2012).

Nelson Llanos is a Master of International Relations and Professor of Contemporary History at Playa Ancha University in Valparaíso, Chile. He is a member of the Latin American Antarctic Historians Association and Director of the Hemispheric and Polar Studies Centre, Viña del Mar, Chile.

Ben Maddison received his doctorate in labour history at the University of Wollongong in 1995. He currently lives on Bruny Island in southern Tasmania. He is a University Associate at the Institute for Marine and Antarctic Studies, University of Tasmania, and a Senior Research Fellow at the University of Wollongong. His research interests include the subaltern histories of Antarctica, working-class and colonial history, and the history of the commons in Australia. He is currently writing a social and political history of the Southern Ocean.

Dr Jeffrey McGee is a Senior Lecturer in Climate Change, Marine and Antarctic Law at the Faculty of Law and the Institute for Marine and Antarctic Studies at the University of Tasmania in Hobart, Australia. He is Director of the Australian Forum for Climate Intervention Governance and a member of the Centre for Marine Socioecology at University of Tasmania. He is an Assistant Editor of the *Carbon and Climate Law Review* and an advisory board member for the Forum for Climate Engineering Assessment at American University, Washington D.C.

Dr Hanne Nielsen is a Lecturer in English at the University of Tasmania in Hobart, Australia. She is on the Executive Committee of the SCAR Standing Committee on Humanities and Social Sciences (SC-HASS); is the book review editor of *The Polar Journal*; and is a past President of the Association of Polar Early Career Scientists (APECS). Hanne spends her summers working as a tour guide in the Antarctic Peninsula, and her winters in Hobart.

Dr Carolyn Philpott is a Senior Lecturer in Musicology at the University of Tasmania's Conservatorium of Music, within the School of Creative Arts and Media. Her research focuses on twentieth-century music and on intersections between music, place and the environment. She has published numerous articles on Antarctic-related music in scholarly journals, including in *Organised Sound, Context: Journal of Music Research, Australian Historical Studies, The Polar Journal* and *Polar Record*. She is currently co-editing (with Matt Delbridge and Elizabeth Leane) the volume *Performing Ice* for Palgrave Macmillan's *Performing Landscapes* series.

Dr Juan Francisco Salazar is an Associate Professor in the School of Humanities and Communication Arts and Research Director of the Institute for Culture and Society at Western Sydney University. He is a member of the Standing Committee for Humanities and Social Sciences of the Scientific Committee for Antarctic Research.

Dr Tim Stephens is Professor of International Law and Australian Research Council Future Fellow at the University of Sydney Law School, University of Sydney, Australia. His Future Fellowship research project is examining the implications of the Anthropocene for international law.

Foreword

This extremely rich collection of meticulously researched chapters, and its publication under the imprint of the prestigious *Routledge Environmental Humanities Series*, witness the fact that humanities, law and social science scholarship on Antarctica and the Southern Ocean has come of age. The discursive journey of Antarctica – as a 'Continent of Science and Peace' in the 1950s and 1960s, then through the lenses of global resource geopolitics in the 1970s and 1980s – has now entered a complex and compelling phase of environmental protection. Moreover, the dawning of the new geological epoch of the Anthropocene has placed Antarctica at the centre of both ethical and geopolitical dimensions of climate change.

These diverse representations of Antarctica – and the policies and practices that are simultaneously fed by, and feed into, these constructions – seem to overlap at this juncture in an entangled, intricate pattern. Whereas scientific and peaceful uses remain the two major conceptual and policy pillars of Antarctic governance, the nature and scope of both have multiplied over the years. This volume illuminates various ways of approaching, imagining, visualising and analysing Antarctica today. Also, compressed in the apparently 'exceptional' polar attributes are multiple meanings, memories and values, competing with each other at times for greater salience, authority and even legitimacy. Conceptualising contemporary Antarctica and its 'governance' and mapping the wide-ranging implications of its fast-multiplying human uses – from peaceful-scientific to commercial, such as tourism and bioprospecting – through the deployment of a single disciplinary lens (from either the social or natural sciences) is neither feasible nor desirable. The much-needed intervention by perspectives from the humanities, law and social sciences in this book reveals Antarctica in ways hitherto unimagined and unexplored.

The overarching, and increasingly overwhelming, concern that binds this diverse interdisciplinary collection is the challenge of sustainable *futures* for Antarctica in the Anthropocene. There is a growing sense of urgency based on the acknowledgement that the extent to which human actions are causing far-reaching, and in some cases irreversible, global change to the life-sustaining resources of the blue planet calls for radically different philosophical, social, economic and political views of our environment at various scales.

This new geological epoch is likely to act as a catalyst for novel geopolitical, legal and ethical imaginings of Antarctica and the Southern Ocean. It will also have a significant bearing on the question of what is being threatened by climate change and thus needs to be secured: instruments of governance, the status quo in terms of knowledge–power hierarchy, values enshrined in the Antarctic Treaty, national interests and alignments or colonial geographies of various territorial claims and rights? As the Antarctic Ice Sheet melts and the ground beneath the feet of millions of people, especially in the Global South, literally starts cracking due to global warming in an unequal world, the current narratives centred on the question of climate mitigation in the Antarctic Treaty area will most likely be tempered with normative considerations of fair access and equity with regard to the power-capability generating endowments of Antarctica and the Southern Ocean, including its resources. Whereas a geopolitical economy perspective invites attention to the problematic of *who gets what, when, where and how from a given resource endowment*, the complex human-social geography of the Anthropocene adds what I have described elsewhere (Chaturvedi 2018) as a 'moral economy' perspective to this equation, by pointing out that there is a profound difference between self-proclaimed geopolitical-legal access and a 'legitimate' entitlement to a resource base.

The future prospects of a moral economy in the Antarctic remain entangled with the geopolitics of territorial claims and counter-claims on the Antarctic continent, which have proved quite resilient to the legal 'freezing' under Article IV of the Antarctic Treaty. It looks like 'nationalisms' *in* and *about* the Antarctic will continue to be asserted in diverse sites – even, as Alan Hemmings's chapter in this book argues, in Antarctic ice-cores. At the same time, the universalism enshrined in its Preamble will continue to remind the twenty-nine consultative parties to the Antarctic Treaty, including the twelve original signatories, that the rights they enjoy by virtue of this privileged status come with corresponding duties. The rights–duties interface becomes extremely important, given the imperatives of the Anthropocene. Yet, there is no denying the fact that, despite a global civil society in the making, most of the political imaginations, engagements, academic writings and intellectual conversations about the Antarctic remain locked in what John Agnew (1994) so aptly calls a 'territorial trap'. Nonetheless, a 'sense of responsibility' arises due to values enshrined in various Antarctic instruments, such as 'open access' for science, sharing of knowledge and information and 'peaceful' uses of the space-place in the best interests of humankind. As Swami Vivekananda, a well-known intellectual and philosopher from India, once said: 'All differences in this world are of degree and not of kind, because oneness is the secret of everything'.

Seen through the lens of a critical social science perspective, it becomes clear that a normative appeal for the relentless interrogation of a 'scramble' for resources, the exaggeration of threats, the simplification of complex challenges and the marginalisation of the voices and concerns of those (largely from the Global South) who are not represented in the narratives and practices of Antarctic governance lies at the heart of post-colonial engagement with Antarctica. As the challenge of the Anthropocene unfolds in the decades ahead, the Antarctic Treaty System (ATS) will be compelled to speak on behalf of an 'environmentalism of the poor' which insists on broadening and deepening our understanding of climate change. This will involve looking at data related to global warming and unsustainability on both land and sea, the causes of which are linked to deeply entrenched ecological irrationalities and the relentless pursuit of economic growth-oriented models of 'development'. Once approached through the lens of the Anthropocene, a connection between the slum dwellers of Dharavi in Mumbai (the coastal financial capital of India) and the displacements likely to be caused by incremental sea-level rise due to melting East Antarctic glaciers become apparent.

The global climate is changing, albeit with a complex geography underlying both the causes and consequences. What does not appear to be changing, however, is the 'collective' behaviour of societies, states and regimes. It is not a question of if, but rather when, the evolving climate change narratives at the Antarctic Treaty Consultative Meetings (ATCMs) will be critiqued and questioned by a critical geopolitics of the Anthropocene. Such a critique interrogates the limited nature and limiting scope of these narratives in terms of both ethical and geopolitical considerations around questions of knowledge, values, representation, responsibility and accountability. Climate change carries profound physical as well as ideational implications for the Antarctic, its pronounced/proclaimed exceptionalism and for the legal–political–ethical boundaries of its governance. Powerful visualisations of Antarctica and its ecosystems at the 'receiving end' of climate change, with far-reaching regional and global implications, have so far failed to catalyse individual and collective behavioural change. After all, Antarctic Treaty states include some of the biggest polluters of the atmosphere. Can the current Antarctic climate discourse be broadened, deepened and reconfigured to give voice to the global periphery, especially in the Global South? Those looking for answers to such questions are likely to find guidance in this exceptionally insightful and useful book.

When I went on the Antarctic pilgrimage in the early 1990s, I was awestruck by both the sheer beauty and the extraordinary intellectual depth of the place. It was a deeply moving spiritual experience, revealing the invisible 'sacred' nature of the apparently profane geographies of Antarctica and the Southern Ocean. It did not take long for the realisation to dawn on me that a continent larger than China and India put together was constituted of both tangible materiality and intangible meanings and values. No doubt, pluralising the essence and meaning of Antarctica as a place runs the risk of fragmentation. But as each and every contribution to this volume conveys forcefully in its own vein, the risk is worth taking and, in fact, highly rewarding.

Climate change operates within a climate of fear. Yet, even as geographies of fear are invoked and widely circulated, Antarctica should remain a *Continent of Hope*. Fear of engaging with normative and ethical aspects of the Anthropocene could be jettisoned in the ATS. This is not to suggest that we should bring controversies and contestations from the rest of the world to the Antarctic. Rather, it is an appeal to rethink the concepts of security, sovereignty and sustainability through collective behavioural change *with hope*. The following excerpt from *Gitanjali*, for which the Indian poet Rabindra Nath Tagore was awarded the Nobel Prize for Literature in 1913, questions the politics of fear. It shows how insightful perspectives from the humanities can be for those in search of alternative imaginings of ecologically sustainable and socially just futures for humankind in the Southern polar region and beyond, based on a progressive sense of place.

> Where the mind is without fear and the head is held high
> Where knowledge is free
> Where the world has not been broken up into fragments
> By narrow domestic walls
> Where words come out from the depth of truth
> Where tireless striving stretches its arms towards perfection
> Where the clear stream of reason has not lost its way
> Into the dreary desert sand of dead habit
> Where the mind is led forward by thee
> Into ever-widening thought and action
> Into that heaven of freedom, my Father,
> let my country [humanity] awake

Sanjay Chaturvedi
South Asian University
India

References

Agnew, J 1994, 'The territorial trap: the geographical assumptions of international relations theory', *Review of International Political Economy*, vol. 1, pp. 53–80.

Chaturvedi, S 2018, 'The future of Antarctica: minerals, bioprospecting and fisheries' in M Nuttall, TR Christensen & MJ Siegert (eds), *The Routledge Handbook of the Polar Regions*, Routledge, New York, pp. 403–415.

1 Anthropocene Antarctica

Approaches, issues and debates

Elizabeth Leane and Jeffrey McGee

The Antarctic is a region that traditionally occupied the remote reaches of the geographical imagination. In the Anthropocene, however, the 'frozen continent' has become central to the planet's present and future. Even as ice cores taken from its interior reveal the deep environmental history of the planet, warming ocean currents are ominously destabilising the glaciers around its edges. The continent contains over ninety per cent of the world's ice, with the potential to raise sea levels by nearly sixty metres, if it were all to melt. While such a wholesale melt of the Antarctic ice sheet is not imminent, estimates (based on a business-as-usual greenhouse gas emissions scenario) indicate the continent's ice could contribute over a metre of sea-level rise by the end of this century and over fifteen metres by 2500 (DeConto & Pollard 2016). And warming global average temperature – along with associated effects, such as ocean acidification and species migration – are only some the hallmarks of the global-scale threats to the region's environment arising from activities remote from the continent itself. Marine microplastics pollution, possibly originating from outside the region, has been found in Antarctic waters (Waller et al. 2017). The thinning of the ozone layer in the atmosphere above the continent, identified by Antarctic scientists in the 1980s, has begun to abate due to international action to reduce the use of ozone-depleting gases, but recovery of ozone concentration to 1980s levels is not expected until the second half of this century (World Meteorological Organization 2018, p. 3). For many decades framed as a 'last wilderness', Antarctica is now increasingly understood as an environment irrevocably altered by remote human action and one that will irrevocably change the course of human lives all over the globe.

Scientific research into the impact of these global environmental changes on the Antarctic region is therefore crucial and has become increasingly prominent outside the polar research community. The continent's spectacular icescapes and charismatic wildlife frequently feature on the cover of leading scientific journals such as *Nature* and transfer readily into popular forums. Popular culture has certainly registered the important links between Antarctica and global environmental change, as evidenced by Hollywood films such as the disaster movie *The Day After Tomorrow* (2004) and the children's animation *Happy Feet Two* (2011). Large calving icebergs that might formerly have been enjoyed as spectacular natural occurences are now treated as political events and presaged by news headlines around the world.

The Anthropocene, then, not only brings attention to Antarctica's integral role in the global climate system; it also asks us to rethink our assumptions about a place 'often depicted as paradigmatically non-human' (Roberts, Howkins & van der Watt, 2016, p. 2). Labelled variously a 'continent for science', a 'giant laboratory' and a 'last wilderness', Antarctica in the popular imagination has until very recently been largely thought of as a pristine environment defined by its remoteness from human settlement. But the global biophysical impacts of the Anthropocene significantly challenge the notion of Antarctica as a remote place, cut off from the planet by the circumpolar current and extreme weather of the Southern Ocean. In the Anthropocene, Antarctica becomes nearer in thought, entangled with our everyday actions. If the Anthropocene signals (in geographer Jamie Lorimer's words) 'the end of the idea of Nature as a pure place untouched by human hands' (2017, p. 121), then the Antarctic, previously understood as the material embodiment of this ideal, faces a radical reframing.

The concept of the Anthropocene also brings home the fact that humans never encountered a truly 'pure' Antarctica. When European explorers caught sight of its icy coasts in the early nineteenth century, the gasses in its ice were already recording the atmospheric changes brought about by the Industrial Revolution (British Antarctic Survey 2014). Antarctica and the subantarctic islands were the site of a massive sealing and whaling industry in the nineteenth century, with fur seals driven close to extinction. Explorers who stoically battled the Antarctic elements for national and scientific priorities were also interested, in some cases, in the potential for exploiting the continent's mineral resources.[1] In the same year the Antarctic Treaty entered into force (1961), the United States installed a nuclear reactor to power McMurdo Station, its largest research base. The Treaty designated Antarctica a 'continent for science and peace', but the decades that followed also saw krill, icefish and later toothfish become the subject of industrial fishing in the region, and whales being hunted by Japanese factory ships. And although the 1991 Protocol on Environmental Protection to the Antarctic Treaty (Madrid Protocol) introduced a series of measures to limit the environmental impact of activities in Antarctica, including the banning of non-native species (bar humans), the human 'footprint' on Antarctica is not trivial and cannot be blamed solely (or even mostly) on the burgeoning tourist industry. While over fifty thousand tourists visited the Antarctic region in the 2017–2018 summer season – some of them perhaps motivated by an Anthropocenic 'last chance to see' impulse – research suggests that scientific programmes have a greater environmental impact,[2] with infrastructure and related disturbance 'similar in size to the total ice-free area of Antarctica … [and] disproportionately concentrated in some of the most sensitive environments' (Brooks et al. 2019, p. 185).[3] In 2017, there were seventy-six research stations on the continent (COMNAP 2019).

While these kinds of impacts are 'local' rather than global problems, in the sense that they are a result of human activity in the region, they nonetheless form part of a new understanding of Antarctica as an anthropogenic icescape, enmeshed with the human world in ways that require ongoing analysis, critique and reassessment.

This creates a challenge for scholarship, as the disciplines designed to think critically about human culture and society have only recently begun to apply their conceptual frameworks to Antarctica. Before turning to this topic, however, it is necessary to tease out the multiple and contested uses of the term 'Anthropocene', particularly as they relate to the Antarctic region.

Antarctica in the Anthropo-scene

In this volume we understand 'the Anthropocene' to broadly indicate the period in which human activity of various kinds has become a key driving force of planetary environmental change. In the last two decades, the concept has gained considerable traction in scholarship across a wide range of disciplines. The exact scientific definition of the term, and indeed its acceptance as a geological unit, is far from settled, with the Anthropocene Working Group (AWG) still in the process of preparing an official proposal to the appropriate subset of the International Commission on Stratigraphy (Zalasiewicz et al. 2017, p. 59). Although the scholarly community eagerly anticipates the results of this proposal submission, the concept has gained an interdisciplinary currency which seems unlikely to be diminished by any lack of scientific consensus. The idea of the Anthropocene has been so productive within a wide range of disciplines that it now constitutes its own subfield, with corresponding conferences and journals[4] – an 'Anthropo-scene', as geographer Jamie Lorimer (among others) has described this 'event-space', which now extends beyond academia to the media and other parts of the public sphere (2017, p. 118). As Lorimer writes in an overview of relevant scholarship, the term has 'proliferated promiscuously in ways unforeseen by its creators', and while future acceptance by geoscientists might give it a 'scientific legitimacy', this 'matter[s] less than the pressing problems the Anthropocene names' (p. 131).

Within the humanities, law and social science disciplines (henceforth HLSS), debates about the Anthropocene and the length of the period it describes are themselves productive, generating new insights. Criticisms of the term, often made alongside suggested alternative namings and framings, have centred on its homogenising power to disguise global inequities, presenting as it does 'humanity', rather than certain sectors of it, as the source of destructive planetary-wide environmental change. Also at issue is its apparent reinstatement of human ascendency, 'conceptually hardening modern humanity's perceived entitlements … enshrining humanity's domination over the planet' when exactly the opposite conceptual move is required (Crist 2007, p. 52). Related debates about the date of the Anthropocene's beginning have political meaning, signalling a different root cause of the planet's current trouble: a suggested starting date of the onset of the Industrial Revolution, for example, ties the concept to the emergence of capitalism, whereas another suggestion – the beginning of European settlement of the Americas – points towards Western colonial expansion (Lorimer 2017, p. 132). The AWG is currently proposing the mid-twentieth century as the most likely starting point of the Anthropocene, but the question remains to be decided

(Zalasiewicz et al. 2017, p. 58). We have therefore imposed no restrictions here on how contributors to this volume understand the Anthropocene, leaving them to make their own implicit or explicit case for interpretation, and to embrace or question the term as they see fit.

Within the huge body of HLSS-based scholarship on the Anthropocene, the Antarctic region's scientific relevance – as an 'icy archive' of greenhouse gas data, a global climate 'canary' and a source of massive sea-level rise – is taken as read. Much less explored are the distinct opportunities that the continent currently offers as a subject of social and cultural enquiry. The chapters in this volume bear out this claim, but here we offer a few broad ways in which Antarctica presents a useful and distinctive place – intellectually speaking – from which to address some of the challenges of the Anthropocene.

The growth of the Anthropocene as a subject of scholarly discussion has been accompanied by an increasing interdisciplinarity in academic exchange. Within the environmental humanities, critics such as Noel Castree have urged researchers to 'get their hands dirty in the places [where] scientists operate' and seek 'institutional and epistemological forms of engagement that might alter important conversations occurring outside the humanities' (2014, p. 244). The Antarctic scholarly community is unusually well placed to initiate and maintain these kinds of interdisciplinary engagements. With Antarctica long labelled the 'continent for science', researchers in the HLSS disciplines interested in the region have had to carve out a place for themselves within a science-dominated field – a situation that has generated opportunities as well as challenges. The key body representing Antarctic-based HLSS scholarship is the Standing Committee on Humanities and Social Sciences (SC-HASS) within the international Scientific Committee on Antarctic Research (SCAR).[5] SC-HASS holds its own regular conferences and forms a significant part of SCAR's large biennial Open Science Conferences (OSCs). These represent truly multi-disciplinary events, where researchers in fields ranging from astronomy to literary studies meet and take part. And given SCAR's role in providing advice to the Antarctic Treaty Consultative Meetings, and its connections with industry bodies such as the International Association of Antarctica Tour Operators, the opportunity for transdisciplinary exchange at these meetings is also significant.

This is not to suggest that Antarctic research is a paragon of disciplinary collaboration and inclusion. Authority is still heavily weighted towards the natural sciences and it is not unusual in Antarctic forums for scientists to be positioned as experts on topics that are the traditional purview of the HLSS disciplines (although the opposite situation is extremely rare). For example, at a SCAR OSC held in mid-2018, a well-attended Q&A panel event focused on 'Polar Change and the Future of Society'. Sponsored by the journal *Nature*, the panel comprised five scientists (with another chairing), the head of the UK's Polar Regions Department and one of the journal's senior editors. Given the large number of social scientists presenting at the conference, their absence from the panel drew some comment.[6] For genuine intellectual exchange, HLSS researchers need a seat at this kind of table, and as well as the freedom to challenge the idea of the natural sciences as the altruistic generator of all rele-vant knowledge on Antarctica. Nonetheless, the continual growth and inte-gration of HLSS into the Antarctic research community suggests that, in the future, this will be a particularly active space for disciplinary conversations.

Another striking aspect of Antarctica in the context of a global environ-mental crisis is its status as an internationally governed space. The Antarctic Treaty, which puts the seven territorial claims on the continent into a form of indefinite hiatus – effectively on hold, but not abandoned – has now been in place for sixty years. Long the subject of literary fantasies about alternative societies that invert and rectify the injustices of the wider world, in the twenty-first century Antarctic governance continues to be seen in a utopian light, as presenting an example of international cooperation for the global 'common good' and environmental protection that might provide a model for future governance of other areas, especially outer space. Perhaps, this thinking goes, a reverse globalisation is possible, where humanity's international governance of Antarctica, the erstwhile exception, sets the standard for the rest of human activities. Again, this harmonious image is in many ways naïve. Claimant states have never ceded their territories and, as Alan Hemmings argues in this volume, nationalist motives can influence seemingly noble scientific ventures. Criticisms directed at the Anthropocene as a homogenising concept that smooths over global inequities can equally be directed at the Antarctic Treaty System (ATS) itself. Nations such as Malaysia have protested at various points, particularly during the 1980s, about the perceived dominance of Antarctic governance by Western nations that are wealthy enough to sustain scientific programmes, instead calling for the region to come under UN governance. While some commentators celebrate the longevity and stability of the ATS, others (including Tim Stephens in this volume) point to the mounting pres-sures on the governance regime – such as increased access and a growing diversity of states involved in its governance – that the Treaty was not specifi-cally designed to address (see also Hemmings 2017, p. 518). Despite these dif-ficulties, however, Antarctica still functions not only as a 'laboratory for science' but, as Juan Francisco Salazar suggests in this volume, 'also a laboratory for thinking alternative ways of living in the Anthropocene'.

Lastly, in an 'Age of Humans' the Antarctic offers an ironic but productive point from which to investigate human relationships with the nonhuman world. As we discuss below, critics challenge the rhetorical emptying-out of a continent that sees yearly visits from five thousand station-based workers of various kinds and over fifty thousand tourists (not to mention ships' crews and tour expedition staff). Nonetheless, the Antarctic *is* distinct from other continents: the first recorded human landing was two centuries ago and, although people have been born in and made homes on the continent, no one lives there permanently. Antarctica thus raises a series of philosophical questions that are particularly pertinent at a time when the habitation and exploitation of outer space is being touted as a possible long-term response to global environmental degradation. What are our moral and ethical obligations to an environment that has no indigenous or permanent human population? How should we inhabit a place where we do not traditionally belong – if we should inhabit it at all? What human activities should be considered legitimate in such a region? Should science take precedence over art or sightseeing? What modes of place-making have characterised our relationship with the Antarctic continent thus far? What human anxieties does such an environment generate? Should such a place be designated exceptional so as to protect it, or does such a discourse only serve to disguise the degree to which humans have already impacted it? Researchers in the HLSS disciplines have only recently begun to address these questions.

Antarctica and the HLSS disciplines

Until the years leading up to the turn of the twenty-first century, Antarctica only marginally entered into the field of view of the HLSS disciplines. While creative artists have produced responses to the continent for centuries, first working within a natural history tradition as members of exploratory expeditions, and later via residencies funded and managed by national programmes, researchers from the non-scientific disciplines have been far more reluctant to venture intellectually into the far southern latitudes (and have had fewer opportunities to travel there physically). This is truer of some disciplines than others. The unusual legal status of Antarctica, a region claimed by seven states but functioning as an international space governed by a unique treaty, has interested legal scholars for several decades (see e.g. Auburn 1982; Triggs 1986; Joyner 1998). Debate around environmental protection, tourism and heritage has also attracted researchers in the social sciences, largely coming from management and policy-related perspectives.

The humanities, however, have been far slower in their engagement with the continent. Antarctica's lack of an indigenous or permanent human population, together with an exceptionalist mindset which framed the continent as immune to the political, social and economic forces that affect the rest of the globe, might have made it seem off-limits to analysis outside an accepted natural-science framework. Only in the last decade has the subfield of humanities-based Antarctic research reached a critical mass and a confidence in its own identity, marked by the publication in 2016 of the collection, *Antarctica and the Humanities*. While researchers in the humanities, as well as the social sciences and law, are now thinking through humans' changing relationship with the nonhuman world of the far south, there is far more to be said. It is the humanities, after all, that examine (to use Helen Small's formulation) the 'meaning-making practices of human culture, past and present' (2013, p. 23). While it is easy to think of Antarctica as being devoid of culture, in fact this is a place that the vast majority of the human population only ever experiences through texts – aural, oral, written and visual. The images that we make and stories that we tell about the past, present and future of Antarctica reveal as much about our relationship to the place, as do our behaviour, our international obligations and our domestic laws.

As mentioned above, one of the current debates within the HLSS community engaged with Antarctica is around the notion of 'Antarctic exceptionalism'. This exceptionalism 'creates a separate sphere for Antarctica', emphasising its lack of inhabitants, conflicts and history – an attempt to maintain the continent as a 'blank' space in human terms (Glasberg 2012, p. 96). Researchers in the HLSS disciplines are increasingly demonstrating the dangers of unthinking exceptionalism, as shown in a number of the chapters in this volume.[7]

This challenge to exceptionalist approaches might raise the question of why a volume such as this is singles out Antarctica as an object of social, cultural, historical and legal analysis in the Anthropocene, rather than contextualising it within a broader, planetary category, such as the 'polar regions' or the 'cryosphere' (the snow and ice regions of the Earth). Certainly, current criticism focused on the wider cryosphere is directly relevant to the chapters collected here (see Gough 2013; Sörlin 2015a; Dodds 2018). Environmental historian Sverker Sörlin (2015b, pp. 95, 97) has argued for the emergence of an 'Arctic Humanities' which, like the environmental humanities, is born out of 'crisis' and, like the environmental humanities, must face 'outward, to the tropics, the oceans, the deserts the plains and the cities', thus 'work [ing in] lock step with the humanities elsewhere'. Both the Antarctic and Arctic humanities could be seen as part of a broader cultural focus on the entangled meanings and material qualities of ice.[8] Writing in reference to Himalayan glaciers, Marcus Nüsser and Ravi Baghel (2014, p. 150) suggest the idea of 'the cryoscape' as a way of capturing the 'coming together of human epistemic practices with the physical phenomena that constitute the cryosphere'. Given that the concept of landscape, with its 'capacity to shuttle between … different registers, [acting] as a meeting point for imaginative and material worlds' has already been proposed as a key way of understanding the Anthropocene, the 'cryoscape' seems a useful term for current ways of thinking about ice (Matless 2016, p. 118). There might even be an argument for an emergent 'cryo-Humanities'.[9]

Ice, however, does not exhaust the meaning of 'Antarctica', which is a place demarcated formally by legal and geopolitical discourse, historically and socially within accounts of exploration and travel, and affectively in narratives of personal and spiritual encounter. International legal jurisdiction over 'the Antarctic' varies between the relevant treaties and institutions which govern the Antarctic continent and Southern Ocean. The Convention on the Conservation of Antarctic Marine Living Resources (CCAMLR), for example, uses a simplified linear version of the Antarctic Convergence (the circumpolar zone of the Southern Ocean where seawater temperatures drop suddenly). This biophysical boundary might itself move south as the planet warms. The Antarctic Treaty, by contrast, uses a cartographic line, with all land and attached ice lying south of the sixty degrees south latitude line falling within its jurisdiction. The more recent Madrid Protocol on Environmental Protection applies to the same area.

In the context of a warming planet, the Madrid Protocol's protection of the Antarctic region from the exploitation of mineral resources is significant.[10] Various expeditions have detected mineral deposits, including hydrocarbons, both in continental rock and in the surrounding sea bed, although it is unclear whether, or under what circumstances, these might be commercially viable. The Protocol's ban on mining is indefinite, although it can be modified at any time according to the Antarctic Treaty's procedures or reviewed from 2048. While there are no imminent plans to 'mine Antarctica', there is 'a pervasive sense that the question of owning and exploiting the resources Antarctica remains a live one' (Hemmings, Dodds & Roberts 2017, p. 10). The region's key role in global carbon sequestration, as well as its ice, could play a significant part in the future of the planet (as discussed in Jeff McGee's chapter in this volume).

Antarctica's distinct historical and geopolitical status means that it cannot be equated with ice, nor can humanities-based approaches to understanding the continent's meaning function in isolation from the social sciences and law. A central aim of this volume is therefore to bring together diverse voices from across these disciplines to consider human/nonhuman relationships and interactions in the Antarctic region in the Anthropocene.

★

The chapters presented in this book are grouped into three parts. The first, 'Governance and Geopolitics', looks at a number of key challenges which the Anthropocene presents to the Antarctic Treaty System and Antarctic geopolitics. Tim Stephens opens this part by examining the pressures that global anthropogenic forces are placing on the legal governance of the Antarctic region, asking 'What might an ATS that understands and responds to the challenges of the Anthropocene look like?'. Stephens argues that the exceptionalist mentality that has seen the ATS remain 'aloof' from broader global issues requires urgent rethinking and that global climate change provides an important vehicle for this task. For the future of both the continent and the planet, Stephens urges a 'closer integration' between existing global climate change governance and the institutions of Antarctic governance.

Ice cores are an icon of contemporary climate science, and the drilling of these artefacts is frequently presented in the media in a positive light, as a heroic scientific venture undertaken in extreme conditions. In 'Subglacial nationalisms', Alan Hemmings queries the motivation behind the quest for a 'million-year' ice core, detecting in this much-touted activity a nationalism 'disguised' as a contribution to 'international public goods'. The challenges of the Anthropocene, he argues, can readily be put to work in the service of nationalist agendas.

One response to the Anthropocene has been the proposal of planetary-scale technological fixes. In 'Frozen Eden lost', Jeff McGee examines the idea of geoengineering Antarctica and the Southern Ocean to reduce the impacts of human-induced climate change. McGee outlines several actual proposals for large-scale human intervention in the Antarctic cryosphere and Southern Ocean, including increasing the reflectivity of ice-covered areas; enhancing sequestration of carbon in the deep ocean; and designing undersea walls or berms to slow the deterioration of ice sheets. He is primarily interested in the way these current scientific and policy discourses of Antarctic geoengineering are further challenging the sense of separateness between Antarctica and wider world. 'The Anthropocene', he contends, is 'reshaping Antarctica in both a material and a discursive sense'.

The last chapter in this part, Juan Francisco Salazar's 'The Anthropocene melt', takes a more conceptual approach, reflecting on Antarctica as 'an important object through which to think the Anthropocene'. Salazar is particularly interested in working through the relationship between ice and time. Antarctica, he argues, has always encouraged an orientation towards the future – something that the volume of science fiction set there, as well as contemporary scholarship, confirms (Leane 2013; Tin et al. 2014; Liggett et al. 2017). Insights produced from an engagement with the continent's 'geo-logics', Salazar suggests, can be applied on a planetary level.

The second part of *Anthropocene Antarctica* turns from the legal and geopolitical to the textual and the cultural. The idea of the Anthropocene has put ecocriticism under some strain, with critics such as Timothy Clark (2015, p. 21) pointing to 'the limits of cultural representation as a force of change in human affairs'. In this sense, it is perhaps significant that two of these chapters – Elizabeth Leane's and Hanne Nielsen's – focus on texts with high circulation. Leane looks at ecothrillers set in Antarctica, following the way in which the icescape – both as a reflection of the plot and an actor that advances the plot – functions to mediate between the global scale of environmental disaster and the local scale on which the action is played out. She also warns against the temptation to treat the future of Antarctica as though it were part of a thriller plot: the climate crisis has no firm deadline, and is best addressed not through heroic individual deeds in a race against the clock, but rather collective action in the mundane present.

In 'Save the penguins', Nielsen takes advertising images as 'a proxy for accessing dominant attitudes towards the far south'. Focusing particularly on images which emphasise the region's environmental fragility, she traces the ways in which this idea is ironically put to work in the service of selling products and services. David Matless has pointed to the possibilities of certain landscapes – 'coastlines, glacier snouts, ice sheet edges, felled forest and the like' – to provide 'stepping points for the humanities as a teller of Anthroposcenic stories' (2016, p. 118). Nielsen warns of the complexity of using Antarctic spectacles for environmental purposes, pointing out that such images can all too readily function as 'Ice-wash'.

The remaining chapter in this central section, Carolyn Philpott's 'Listening "at the sea ice edge"', examines sonic art works by Douglas Quin and Philip Samartzis. Both artists incorporate into their compositions found sound – anthropogenic and environmental – recorded during their journeys to Antarctica, bringing attention to the multi-sensorial quality of the icescape and surrounding water. Such Antarctic soundscape compositions, Philpott argues, are not only the preserve of a small coterie of enthusiasts, but are encountered in exhibitions and museums, on film and television, as well as being available as recordings, enabling a wide range of people to sonically experience an environment that they are unlikely to encounter directly. These compositions have the ability to increase awareness and appreciation of the Antarctic environment, without the carbon footprint of travel or the local impact of landing on the continent. The soundscape compositions are also very much of the Anthropocene in the sense that the human artist is no longer a sole creator, but rather working in collaboration with the nonhuman world.

The last part of the volume focuses on the question of human presence in Antarctica, thinking through questions of inhabitation, settlement and place in a continent renowned for its lack of humans. Like the idea of the Anthropocene itself, the tendency to erase human presence from Antarctica masks a more invidious absence of certain kinds of people – both from the continent itself and from accounts of its history. Researchers such as Morgan Seag, Meredith Nash, Lize-Marié van der Watt and Ben Maddison have examined the ways in which the operation of gender, race and class politics has determined who travels to Antarctica and why, and whose stories are repeated upon return (see e.g. Maddison 2014; van der Watt 2016; Nash et al. 2019; Seag 2019). In this volume, Maddison turns to questions of indigeneity that have for too long been ignored in relation to Antarctica. While Maddison points to traditions of Antarctic travel and connection in various indigenous groups in far southern locations, his primary focus is the use of indigenous people and technologies in the 'Heroic Era' of exploratory ventures. 'The assumption has been made', he writes, 'that because … there were no indigenous "Antarcticans", and because Antarctic exploration was overwhelmingly undertaken by Europeans, indigeneity could not be relevant to understanding its history'. His chapter demonstrates the falseness of this assumption and points to the need for further research on the relationships between indigeneity and the Antarctic region.

People are often surprised to learn that humans have been born in Antarctica, that families have lived there, and that children have been schooled there. Nelson Llanos provides a detailed account of this under-examined aspect of Antarctic history, drawing directly from interviews with inhabitants of Chile's Villa Las Estrellas during the 1980s. Llanos explores the political context of Chile's decision to people the base with military families, as well as the challenges faced by the inhabitants of this Antarctic town – particularly the 'housewives' – and the strategies they drew on to deal with their unusual geographic and social situation. Reminding us that Antarctica's history is a lot more complex and varied than is often assumed, Llanos argues for greater attention to 'the social aspects of human presence in the icy continent', including 'the history of the family, children and women, and all people who have remained outside of the history of Antarctica'. This chapter reminds us of the various scales of human–nature interaction with Antarctica, with the local and personal stories recounted here contrasting with the global questions of other chapters.

The final chapter in this part is historian Adrian Howkins' 'Placing the past'. The way Antarctica relates to the concept of place has been a topic of interest in the last few years (see e.g. Antonello 2016). Against the universalising tendencies of the Anthropocene, Howkins draws on the example of the

McMurdo Dry Valleys to make a case for local specificity, both in scientific and historical research. Howkins cautions against an overly homogenising approach to the history of particular places in Antarctica, instead calling for a 'focus on embracing both distinctiveness and connectivity'. Howkins' chapter suggests that Antarctica, in its very resistance to conventions of traditional notions of place, might be a productive site from which to interrogate the meaning of this concept in the Anthropocene.

This volume is offered in the spirit of further opening up conservation and analysis about human–nonhuman interactions in Antarctica and the Southern Ocean. The chapters that follow draw from a wide range of disciplines including literature, law, geopolitics, musicology and cultural history. What holds this collection together is our collective view of the Anthropocene as an important lens that can assist us in teasing out these interactions in new and interesting ways. We therefore hope that you will find this volume a rich resource for this task that helps us better understand Antarctica's past and present, as well as its possible futures.

Notes

1 Australian geologist Douglas Mawson made no secret of the potential worth to his country and empire of what he termed the 'commercial resources' of Antarctica, which in his speculations included minerals, wind-energy and 'sightseeing', as well as more unexpected possibilities such as fox-farming and sanatoria. See Mawson 2010, pp. 213–216.
2 A 2009 comparison of tourist versus national programme impact in terms of person days suggests the latter is larger by well over an order of magnitude (Jabour 2009, p. 225).
3 This quotation is from a recent assessment of the footprint of national programmes.
4 Since 2013, Elsevier has published *Anthropocene*, an 'interdisciplinary peer-reviewed journal that addresses the nature, scale and extent of interactions between people and Earth processes and systems'. See <https://www.journals.elsevier.com/anthropocene>. SAGE's *Anthropocene Review* is 'a trans-disciplinary journal … on all aspects of research pertaining to the Anthropocene', its first issue having appeared in 2014. See <https://journals.sagepub.com/home/anr>.
5 Information about the group can be found at <https://www.scar.org/science/hass/sc-hass/>.
6 A video of the event can be found at <https://vimeo.com/304567022>, with the Q&A panel beginning about twenty-five minutes in. The issue of the scientific make-up of the panel comes up towards the end of the event.
7 This not, however, a wholesale rejection of exceptionalist thinking. Hemmings, for example, argues that a 'new deliberate exceptionalism', replacing the 'reflexive exceptionalism we once had', is needed to protect the region from 'global norms' that are 'problematical in Antarctica because they have evolved in different parts of the world' (Hemmings 2009, p. 71).
8 Examples include Sörlin 2015a; Bjørst 2010; Glasberg 2011; Antonello 2017; and Dodds 2018.
9 Geopolitics scholar Klaus Dodds has begun referring to an 'ice humanities' closely associated with the 'blue humanities'. See Dodds's profile on the *Marine Policy* journal website (Elsevier 2019).
10 The Protocol came only after extensive discussion in the 1980s about the need for a convention on mining. Indeed, such an instrument – the Convention on the Regulation of Antarctic Mineral Resource Activities – was signed by numerous states in 1988, but never entered into force. The Madrid Protocol effectively replaced it.

References

Antonello, A 2016, 'Finding place in Antarctica' in P Roberts, L-M van der Watt & A Howkins (eds), *Antarctica and the humanities*, Palgrave Macmillan, London, pp. 181–203.

Antonello, A 2017, 'Engaging and narrating the Antarctic ice sheet: a history of an earthly body', *Environmental History*, vol. 22, pp. 77–100.

Auburn, FM 1982, *Antarctic law and politics*, C Hurst, London, and Croom-Helm, Canberra.

Bjørst, LR 2010, 'The tip of the iceberg: ice as a non-human actor in the climate change debate', *Études/Inuit/Studies*, vol. 34, no. 1, pp. 133–150.

British Antarctic Survey 2014, 'Ice cores and climate change', viewed 25 March 2019, <https://www.bas.ac.uk/data/our-data/publication/ice-cores-and-climate-change/>.

Brooks, ST, Jabour, J, Van der Hoff, J & Bergstrom, DM 2019, 'Our footprint on Antarctica competes with nature for rare ice-free land', *Nature Sustainability*, vol. 2, pp. 185–190.

Castree, N 2014, 'The Anthropocene and the environmental humanities: extending the conversation', *Environmental Humanities*, vol. 5, pp. 233–260.

Clark, T, 2015, *Ecocriticism on the edge: the Anthropocene as a threshold concept*, Bloomsbury, London.

COMNAP (Council of Management of National Antarctic Programs) 2017, *Antarctic Station Catalogue*, COMNAP Secretariat, Christchurch, viewed 20 March 2019, <http s://www.comnap.aq/Members/Shared%20Documents/COMNAP_Antarctic_Sta tion_Catalogue.pdf>.

Crist, E 2007, 'Beyond the climate crisis: a critique of climate change discourse', *Telos*, vol. 14, pp. 29–55.

DeConto, RM & Pollard, D 2016, 'Contribution of Antarctica to past and future sea-level rise', *Nature*, vol. 531, pp. 591–597.

Dodds, K 2018, *Ice: nature and culture*, Reaktion, London.

Elsevier 2019, 'Professor Klaus Dodds, PhD. Editorial Board, *Marine Policy*', viewed 25 March 2019, <https://www.journals.elsevier.com/marine-policy/editorial-board/p rofessor-klaus-dodds-phd>.

Emmerich, R (dir) 2004, *The day after tomorrow*, motion picture, 20th Century Fox, Los Angeles.

Glasberg, E 2011, 'Living ice: rediscovery of the poles in an era of climate crisis', *WSQ: Women's Studies Quarterly*, vol. 39, no. 3–4, pp. 221–246.

Glasberg, E 2012, *Antarctica as cultural critique: the gendered politics of scientific exploration and climate change*, Palgrave, New York.

Gough, R (ed.) 2013, 'On ice', *Performance Research: Special Issue*, vol. 18, no. 6.

Hemmings, AD 2009, 'From the new geopolitics of resources to nanotechnology: emerging challenges of globalism in Antarctica', *Yearbook of Polar Law*, vol. 1, pp. 55–72.

Hemmings, AD 2017, 'Antarctic politics in a transforming global geopolitics' in K Dodds, AD Hemmings & P Roberts (eds), *Handbook on the politics of Antarctica*, Edward Elgar, Cheltenham, pp. 507–522.

Hemmings, AD, Dodds, K & Roberts, P 2017, 'Introduction: the politics of Antarctica' in K Dodds, AD Hemmings & P Roberts (eds), *Handbook on the politics of Antarctica*, Edward Elgar, Cheltenham, pp. 1–17.

Jabour, J 2009, 'National Antarctic programs and their impact on the environment' in RK Knowles & MJ Riddle (eds), *Health of Antarctic wildlife*, Springer, Berlin and Heidelberg, pp. 211–230.

Joyner, C 1998, *Governing the frozen commons: the Antarctic regime and environmental protection*, University of South Carolina Press, Columbia.

Leane, E 2013, 'Yesterday's tomorrows and tomorrow's yesterdays: utopian literary visions of Antarctic futures', *The Polar Journal*, vol. 3, no. 2, pp. 333–347.

Liggett, D, Frame, B, Gilbert, N & Morgan, F 2017, 'Is it all going south? Four future scenarios for Antarctica', *Polar Record*, vol. 53, no. 5, pp. pp. 459–478.

Lorimer, J 2017, 'The Anthropo-scene: a guide for the perplexed', *Social Studies of Science*, vol. 47, no. 1, pp. 117–142.

Maddison, B 2014, *Class and Colonialism in Antarctic Exploration, 1750–1920*, Pickering & Chatto, London.

Matless, D 2016, 'Climate change stories and the Anthroposcenic', *Nature Climate Change*, vol. 6, pp. 118–119.

Mawson, D 2010, 'The commercial resources of Antarctica – (IV) general' in A McLean & D Mawson (eds), *The Adelie Blizzard: Mawson's forgotten newspaper*, Friends of the State Library of South Australia, Adelaide.

Miller, G (dir) 2011, *Happy feet two*, motion picture, Warner Bros, Los Angeles.

Nash, M, Nielsen, HEF, Shaw, J, King, M, Lea, M-A & Bax, N 2019, '"Antarctica just has this hero factor …": gendered barriers to Australian Antarctic research and remote fieldwork', *PLoS ONE*, vol. 14, no. 1, article e0209983, doi: doi:10.1371/journal.pone.0209983.

Nüsser, M & Baghel, R 2014, 'The emergence of the cryoscape: contested narratives of Himalayan glacier dynamics and climate change' in B Schuler (ed.), *Environmental and climate change in South and Southeast Asia: how are local cultures coping?* Brill, Leiden and Boston, pp. 138–156.

'Polar change and the future of society. Q&A discussion, based on recent publications in Nature' 2018, SCAR Open Science Conference, Davos, Switzerland, June 2018, viewed 20 March 2019, <https://vimeo.com/304567022>.

Roberts, P, Howkins, A & van der Watt, L-M 2016, 'Antarctica: a continent for the humanities' in P Roberts, L-M van der Watt & A Howkins (eds), *Antarctica and the humanities*, Palgrave Macmillan, London, pp. 1–23.

Seag, M 2019, 'Women in polar research: a brief history', *The Arctic Institute: Center for Circumpolar Security Studies*, viewed 19 March 2019, <https://www.thearcticinstitute.org/women-polar-research-brief-history/>.

Small, H 2013, *The value of the humanities*, Oxford University Press, Oxford.

Sörlin, S 2015a, 'Cryo-history: narratives of ice and the emerging Arctic humanities' in B Evengård, J Nymand Larsen & Ø Paasche (eds), *The new Arctic*, Springer, Cham, pp. 327–339.

Sörlin, S 2015b, 'The emerging Arctic humanities: a forward-looking post-script', *Journal of Northern Studies*, vol. 9, no. 1, pp. 93–98.

Tin, T, Liggett, D, Maher, PT & Lamers, M (eds) 2014, *Antarctic futures: human engagement with the Antarctic environment*, Springer, Dordrecht.

Triggs, GD 1986, *International law and Australian sovereignty in Antarctica*, Legal Books, Sydney.

van der Watt, L-M, & Swart, S 2016, 'The whiteness of Antarctica: race and South Africa's Antarctic history' in P Roberts, L-M van der Watt & A Howkins (eds), *Antarctica and the humanities*, Palgrave Macmillan, London, pp. 125–156.

Waller, CL, Griffiths, HL, Waluda, CM, Thorpe, SE, Loaiza, I, Moreno, B, Pacherres, CO, Hughes, KA 2017, 'Microplastics in the Antarctic marine system: an emerging area', *Science of the Total Environment*, vol. 598, pp. 220–227.

World Meteorological Organization (WMO) 2018, *Executive Summary: Scientific Assess-ment of Ozone Depletion: 2018*, World Meteorological Organization, Global Ozone Research and Monitoring Project – Report No. 58, Geneva, viewed 20 March 2019, <https://www.esrl.noaa.gov/csd/assessments/ozone/2018/executivesummary.pdf>.

Zalasiewicz, J, Waters, CN, Summerhayes, CP, Wolfe, AP, Barnosky, AD, Cearreta, A, Crutzen, P, Ellis, E, Fairchild, IJ, Gałuszka, A, Haff, P, Hajdas, I, Head, MJ, Ivar do Sul, JA, Jeandel, C, Leinfelder, R, McNeill, JR, Neal, C, Odada, E, Oreskes, N, Steffen, W, Syvitski, J, Vidas, D, Wagreich, M, & Williams, M 2017, 'The Working Group on the Anthropocene: summary of evidence and interim recommendations', *Anthropocene*, vol. 19, pp. 55–60.

Part 1
Governance and geopolitics

2 Governing Antarctica in the Anthropocene

Tim Stephens

Introduction

There is growing interest in the implications of the Anthropocene for Antarctica, from the physical impacts of global environmental change through to the ways in which these could challenge the stability of Antarctica's governance norms and institutions. Despite sharpening scientific focus on Antarctica's Anthropocene futures, there has been limited engagement by the Antarctic Treaty System (ATS) with the legal and policy consequences of a state-change to the Antarctic environment. This chapter examines the implications of the Anthropocene for understanding Antarctica as an internationally regulated domain. It contends that the ATS should be more active and vocal in global regime complexes where decisions are being made that will determine the fate of Antarctica in this new era.

What is the Anthropocene?

Geologists are the gatekeepers of the technical definition of the Anthropocene. The term has been proposed as a new geological epoch, and they have sought to avoid entering policy debates. Instead, for geologists, the key question in dating the new era is empirical: when did human impacts produce 'suitable signals in the stratal record?' (Zalasiewicz, Waters & Head 2017, p. 289). Tasked with determining whether the Anthropocene is truly a new geological time unit and, if so, when it began, the Anthropocene Working Group of the International Commission on Stratigraphy has pinpointed the dawn of the nuclear age as the critical date (Zalasiewicz et al. 2015). It was at that moment, in the mid-twentieth century, when the 'Great Acceleration' (McNeill & Engelke 2014) of human population, industry and energy use commenced and started to leave a global geological imprint (Zalasiewicz, Waters & Head 2017).

Taking this date as the genesis of the Anthropocene, we can see the confluence of the three mainstream meanings of the term: the geological meaning (when global changes are discernible in the stratum); the Earth system science meaning (the disturbance to the Earth's bio-geophysical systems); and the socio-ecological–economic connotation (the understanding that these global environmental changes carry major risks for human civilisation). All three notions are captured in

Owen Gaffney and Will Steffen's 'Anthropocene equation' (2017), in which the variable 'H', representing humanity, has become the dominant force of change to the Earth system in recent time.

The Anthropocene is more than a new label for global environmental changes. It represents a revolution in the human–environment relationship. As Clive Hamilton (2015, p. 34) observes, the global environmental transformation in the Anthropocene 'is now telling us that the modern division of the world into a box marked "Nature" and one marked "Human" is no longer tenable'. It is therefore a post-natural epoch, a 'New Earth' in which human and natural forces are intermixed and inseparable (Nicholson & Jinnah 2016, p. 7), with the Earth now 'an interdependent integrated social-ecological system' (Biermann 2014, p. 16). In this context, distinctions between 'hard' and 'soft' sciences, between glaciology and international relations, cannot be maintained. As Hamilton (2015, p. 35) puts it, in the Anthropocene geophysicists must become social scientists and social scientists must become geophysicists, because the planet is now a closely coupled human/natural system.

Dipesh Chakrabarty argues that, because of this dynamic, the Anthropocene involves the convergence of natural history and human history. This, in turn, requires a new politics that looks beyond social forces, such as capital versus labour, and takes seriously the global environmental changes that 'destabilize conditions ... [and] work like boundary parameters of human existence' (Chakrabarty 2009, p. 218). The Anthropocene therefore marks the transition from humanity as the passive inheritor of global environmental conditions set by forces beyond its control to the trustee of planetary environmental stability.

What does the Anthropocene mean for the way we see Antarctica?

In 1959, the United States constructed Camp Century in the Greenland Ice Sheet with the hope of deploying nuclear ballistic missiles there. Despite the centennial ambitions reflected in its name, the facility was abandoned within a decade. However, the base was never properly decommissioned: it was assumed that Camp Century and its hazardous contents would remain forever entombed in the ice. This turned out to be an unsafe assumption. In 2016, glaciologists discovered that, with the melting of the Greenland Ice Sheet, Camp Century and its chemical and radiological wastes are now being released from the ice (Colgan et al. 2016). The retreat of the cryosphere in Greenland, and globally, is one of the most visible markers of the Anthropocene (Wolff 2013, p. 262). Global ice and snow cover continues to decline, and climate change has already delayed the onset of the next ice age for hundreds of thousands of years (Ganopolski, Winkelmann & Schellnhuber 2016, p. 201). That humanity is transforming the global cryosphere both in the present and in the far distant future clearly signifies the scale of the Anthropocene across space and time.

A well-known feature of observed climate change is the amplification of temperature rises at the poles. The warming impacts in the Arctic, where summer sea ice is in a 'death spiral', have been especially pronounced (Wadhams 2016). An average of two degrees Celsius of global warming above pre-industrial temperatures may be the threshold for the complete loss of Arctic summer sea ice (Notz & Stroeve 2016). Because of these stark physical changes in the Arctic, there is a very strong sense that the Anthropocene has arrived at the North Pole, and Arctic researchers are posing and answering research questions in response to this reality. For instance, in 2014 the US National Research Council released a book-length report, titled *The Arctic in the Anthropocene: Emerging Questions* (National Research Council 2014). At the opposite pole, there has been a belief, or wish, or hope, that Antarctica is more resistant or resilient to the Anthropocene. Perhaps Antarctica is the place that this new era has overlooked?

In one sense, the Anthropocene began to leave an imprint in the Antarctic environment from the mid-twentieth century. Whales and seals were the initial victims of the extensive planetary expansion of human activities that reached even to the Antarctic, and the ecosystem effects of industrial whaling and sealing continue to be felt in the Southern Ocean ecosystem (Chown 2017, p. 524). The Anthropocene is inscribed, increasingly obviously, in the depths and surfaces of Antarctica: the industrial chemicals (Nash 2011) and fossil-fuel by-products (Petrenko et al. 2017) trapped in air bubbles in the ice; the ozone hole that remains etched in the southern sky (Solomon et al. 2016); the concentrations of microplastics that have crossed the Polar Front (Waller et al. 2017); and, above all, the melting ice shelves and ice sheets and the growing extent of ice-free areas.

There is considerable natural variation in Antarctic sea ice extent (Matear et al. 2015), although the recent and abrupt retraction is concerning (Turner & Comiso 2017). Far more troubling is the long-term melting of Antarctic ice shelves and ice sheets. Antarctica's ice is not a fixed feature of the landscape, but a dynamic system of moving ice streams that has been in equilibrium in the current interglacial period. While the picture is complex, there is no substantial doubt that Antarctic ice mass is now being lost as temperatures in the region rise, and ice is assailed by warm air from above and by warm waters from below. The decline in Antarctica's ice is the most enduring dimension of Anthropocene Antarctica, as it involves change of sufficient scale to transform both the continent and the planet.

Recent research by Rob DeConto and David Pollard (2016) suggests that Antarctic ice sheets are highly sensitive to global rises in temperature.[1] The last time CO_2 concentrations were as high as they are today was during the Pliocene, three million years ago. At that time, the sea level was ten to thirty metres higher, there was no West Antarctic Ice Sheet and the East Antarctic Ice Sheet was in retreat. Continued growth in concentrations of CO_2 in the atmosphere will trigger the unstoppable collapse of Antarctica's ice, as the buttressing ice shelves melt and the ice sheets slide into the sea. DeConto has described this process as 'literally remapping how the planet looks from space' (quoted in Tollefson 2016, p. 562).

What does the Anthropocene mean for understanding Antarctica as a managed place?

The Anthropocene challenges the reassuring trope that, on a continental scale, Antarctica is an unchanging place. It also collapses the separation between human activities globally and their impacts on Antarctica. To date, we have mostly imagined humanity's physical power over Antarctica, with states projecting an abstract authority through exploration, map-making, nationalist claims and gestures and the passing of myriad Antarctic laws, many of which have little more than symbolic effect. However, in the Anthropocene the future of Antarctica is now in human hands, with the fate of its mass of ice and snow determined from afar by human-induced changes to the global climate. In the Anthropocene, the imagined mastery of Antarctica has given way to a real dominion.

The ATS has made a major contribution to understanding the Anthropocene because of the privileged position that science occupies in Antarctic governance. As Jessica O'Reilly (2013, p. 385) has argued, in this respect Antarctica is best described not as a *terra nullius* or *terra incognita*, but a *terra clima* that is both a barometer of climate change and one of the key 'geographic epicentres of climate science and scientists'. The growing body of Antarctic scientific research has fed directly into the operation of the ATS, and provided the Antarctic Treaty parties with insights into the importance of Antarctica to the Earth system.

Rarely, however, has the ATS expressly recognised the implications of this research. A notable exception is Recommendation XV-14, adopted by the Antarctic Treaty Consultative Meeting (ATCM) in 1989. Although carrying a tepid title ('Promotion of International Scientific Cooperation: A Declaration'), Recommendation XV-14 (1989) expressed the parties' awareness of 'the role that Antarctica and the Southern Ocean play in interactive physical, chemical and biological processes that regulate the total Earth System'. It noted that, 'The Antarctic region has a high negative radiation budget and so acts as one of the Earth's "refrigerators"'. Furthermore, it drew clear connections between global climate change and Antarctica, observing that, 'The Antarctic ice sheet contains enough water to raise global sea level world-wide some 60 metres' and that, 'Any greenhouse warming which makes even a small change to this volume of ice will have a significant impact on sea level' (Secretariat of the Antarctic Treaty 1989). It should be recalled that this text was adopted three years before the conclusion of the 1992 United Nations Framework Convention on Climate Change (UNFCCC).

Recommendation XV-14 (1989) was agreed almost thirty years ago. In speaking of the 'total Earth System', it constituted an important and high-level recognition of the Earth System concept that has been central to understanding the Anthropocene. There were echoes of this resolution in the Commission for the Conservation of Antarctic Marine Living Resources (CCAMLR's) Resolution 30/XXVIII (2009), which recognised that 'climate change is one of the greatest challenges facing the Southern Ocean' and that the Southern Ocean will continue to warm and acidify, and urged 'increased consideration of climate change impacts in the Southern Ocean to better inform CCAMLR management decisions'. CCAMLR Resolution 30/

XXVIII (2009) also requested that the Chairman of CCAMLR write to the President of the Conference of the Parties (COP) to the United Nations Framework Convention on Climate Change (UNFCCC) to 'express that the CAMLR Commission considers that an effective global response by the UNFCCC is urgently needed to address the change of climate change in order to protect and preserve the Southern Ocean ecosystems and their biodiversity'.

Despite major improvements in the understanding of Antarctica's role in global environmental systems, the legal and policy responses within the ATS to the threats that the Anthropocene poses to Antarctica have been limited. This is primarily because the ATS has enabled the Antarctic to be defined as a distinct environmental domain that can be protected and managed by maintaining this separation. However, Earth System science shows that this separation is based on an outdated scientific understanding. As David Bowman has argued, the Anthropocene involves a 'philosophical rupture' for conventional approaches to protecting wild places (quoted in Gill 2016, n.p.).[2] Commenting on bushfires in Tasmania, Australia, in 2016, that were made more severe by climate change, Bowman noted that in the Anthropocene the belief that 'all a wilderness needed was to be left alone' and that it would be safe within the 'confines of [a] park' no longer holds true (quoted in Gill 2016, n. p.). This insight from a fire ecologist also has application to the 'frozen continent', given the centrality of wilderness protection as an Antarctic value and ideal.

The ATS is imbued with the romantic environmentalism of wilderness. While neither in form nor in substance is Antarctica truly a wilderness area,[3] the idea of Antarctica as a natural reserve is a deeply embedded one. Elizabeth Leane (2016, p. 40) explains in her work on representations of the South Pole that the notion of Antarctica as a pristine wilderness – indeed the 'last wilderness', 'last refuge' and 'last hope' – has been central to depictions of the continent since the rise of global environmental consciousness in the 1980s, and the designation of Antarctica 'as a natural reserve, devoted to peace and science' in the 1991 Environmental Protocol. It is therefore no surprise that a recent survey of university students in Spain and the United States found around three quarters of respondents believed Antarctica was fully protected as a wilderness area (Peden et al. 2016).

The idea of Antarctica as wilderness has gained greater impetus and practical expression through the evolution of a range of Antarctic protected area types, including Antarctic Specially Protected Areas, Antarctic Specially Managed Areas, Historic Sites and Monuments and CCAMLR Marine Protected Areas (MPAs). CCAMLR has also designated a unique climate-change-related category of special area, adopting a conservation measure at its 2016 meeting to establish time-limited special areas for scientific study in newly exposed marine areas, following ice shelf retreat or collapse in the Antarctic Peninsula region (CCAMLR 2016).[4] Climate change was also specifically referenced in Conservation Measure 91–05 (2016) establishing the Ross Sea MPA, which notes that the Ross Sea region 'offer[s] rich opportunities for the study of climate change in the region' and that the 'establishment of CCAMLR MPAs can provide important opportunities to understand the ecosystem impacts of climate change separate from those of fishing' (CCAMLR 2016).[5]

Protected areas are valuable tools for safeguarding ecosystems and for building resilience to global environmental changes. But the Anthropocene undermines the assumption that places of wilderness can simply be left undisturbed in order to be protected. As John Dryzek, Richard Norgaard and David Schlosberg (2013, p. 118) put it, 'In the Anthropocene, the very idea of environmental preservation is compromised, as everything we call wilderness is affected by human actions – even if no human physically enters it'. The fundamental challenge of Anthropocene Antarctica is that it is not behaviour on the continent that is changing Antarctica in any radical and unconstrained way, but rather human activities outside the region and beyond the purview of the Antarctic management regime.

While local threats remain serious, the Anthropocene threatens to wash away the relevance and influence of the grand global bargains achieved to protect spaces and places, with some treaties becoming curios, devoted to preserving a natural world that no longer exists. What, for instance, is the purpose of the 1972 World Heritage Convention, if many of the cultural and natural properties it seeks to protect are eviscerated by climate change (UNESCO, USC & UNEP 2016)? This question needs to be asked in relation to World Heritage Properties, such as the Great Barrier Reef, which are threatened by global environmental change. A similar question can be posed in relation to the ATS and Antarctica.

What might an Antarctic Treaty System best suited for the Anthropocene look like?

Now that the Anthropocene has clearly arrived in the Antarctic, what, if anything, should the ATS do about it? Is the Antarctic governance regime destined only to facilitate the observation of global changes, while being unable to take meaningful action to prevent them? What power, if any, does the ATS have over the forces of the Anthropocene?

An initial observation is that, at its inception, the ATS did seek to respond to one of the technological markers of the Anthropocene. Article 5(1) of the 1959 Antarctic Treaty prohibits nuclear explosions in Antarctica and the disposal of radioactive waste.[6] In addressing the risks associated with nuclear weapons and waste, the Antarctic Treaty can therefore be seen as a response to one of the defining technologies of the Anthropocene.

Moreover, the ATS has been at the forefront of whole-of-system environmental governance. This approach better aligns with the demands of effective environmental management in the Anthropocene in which components of the Earth System cannot be seen in isolation.[7] CCAMLR was one of the first fisheries regimes to adopt a whole-of-marine-ecosystem approach to guide its management of Southern Ocean fisheries (Österblom & Olsson 2017, p. 408), while Article II of the Environmental Protocol committed parties to the 'comprehensive protection of the Antarctic environment and dependent and associated ecosystems' (Secretariat of the Antarctic Treaty 1991).

Furthermore, it cannot be doubted that the ATS has played a pivotal role in supporting climate-relevant scientific research, both on its own motion and with the support of the Scientific Committee on Antarctic Research (SCAR). Key developments in this respect include SCAR's 2009 *Antarctic Climate Change and the Environment (ACCE) Report* (yearly updates of which are presented at the Antarctic Treaty Consultative Meetings (ATCM)) and the 2010 Antarctic Treaty Meeting of Experts (ATME) on Implications of Climate Change for Antarctic Management and Governance (ATME Climate Change Report).

Detecting and responding locally to Earth System change is very high on the Antarctic governance agenda, as seen in the Committee for Environmental Protection's (CEP) 'Climate Change Response Work Programme' (CCRWP), which was established on the recommendation of the ATME Climate Change Report. This provides a mechanism for identifying and revising goals and specific actions by the CEP to support efforts within the ATS to respond to the environmental impacts of a changing climate and the associated implications for the governance and management of Antarctica. At ATCM XL – CEP XX in Beijing, the parties in Decision 1 (2017) agreed to establish the Subsidiary Group of the Committee on Environmental Protection on Climate Change Response to support the implementation of the CCRWP by improving coordination of the CCRWP between members, drafting annual updates of the CCRWP and annual progress reports on the implementation of the CCRWP for the CEP to be drawn from its updates to the ATCM (Secretariat of the Antarctic Treaty 2017).

The ATS has therefore been proactive in detecting and recognising the signals of the Anthropocene, and is seeking to address them in order to manage Antarctic ecosystems. But the larger question is what contribution, if any, can the ATS make to strengthening the global norms and institutions that are essential for protecting Antarctica in the Anthropocene?

Both polar regimes face a fundamental challenge in responding to global environmental changes that are driven by activities outside these regions: the problem of 'regime fit'. In this context, 'fit' describes the extent to which an international regime has geographical and other capacity to address an identified governance challenge. Oran Young (2016, p. 213) notes that climate change is the 'quintessential case in point' of mismatch in regime fit as the 'Arctic Council has neither the authority nor the political influence to take significant steps to come to terms with climate change'. The same can be said of the ATS.

In a 2009 review of the response by the Arctic and Antarctic regimes to climate change, Duncan French and Karen Scott observed that both were characterised by 'an emphasis on scientific research, minimal policy initiatives and hortatory recommendations' (p. 649). Since this time, the Arctic Council (the high-level forum for the eight Arctic states) has become more active and 'played the role of messenger, articulating powerful messages about [climate and other] issues based on evidence collected in the Arctic' (Young 2016, pp. 209, 211). The Council has significantly raised the profile of climate change and other Earth System transformations as issues of acute regional and global concern.

For instance, the Arctic Council's Fairbanks Declaration, adopted in May 2017, recognised that 'activities taking place outside the Arctic region … are the main contributors to climate change effects and pollution in the Arctic … underlining the need for action at all levels' (p. 2). It also noted the 'pressing and increasing need for mitigation and adaption actions', including 'global action to reduce both long-lived greenhouse gases and short-lived climate pollutants' (Arctic Council 2017, p. 3). The Arctic Council has sought to contribute directly to deliberations within the UNFCCC. At the UNFCCC COP19 in Warsaw in 2013, the Arctic Council made a statement to the COP which highlighted that, 'Global emissions of greenhouse gases are resulting in rapid changes in the climate and physical environment of the Arctic with widespread effects for societies and ecosystems around the world', and stated unequivocally that, 'It is clear we must reduce carbon dioxide emissions' (Arctic Council 2013, n.p.).

There are no Antarctic equivalents to these Arctic developments, despite climate change occupying increasing space on the ATCM, CEP and CCAMLR agendas. In contrast to the Arctic Council's Fairbanks Declaration, ATCM Resolution 6 (2015) on 'The role of Antarctica in global climate processes' is far more restrained, noting only that, 'Antarctica plays a crucial role in the global climate system' and recognising that, 'Scientific study of Antarctica is crucial to further inform under-standing of global climate processes and their consequential impacts on the entire Earth system' (ATCM 2015, p. 349). But its operative provisions do little more than recommend additional research to support COP21 of the UNFCCC in Paris in 2015. Moreover, the communication between the ATS and the UNFCCC has generally been limited to providing the UNFCCC with ministerial declarations and scientific studies, such as SCAR's 2009 ACCE Report, rather than urging UNFCCC parties to address climate change.[8]

What accounts for the contrast between the 'Anthropocene activism' of the Arctic Council and the 'Holocene hesitancy' of the Antarctic Treaty System? Part of the explanation lies in the fact that Arctic Council declarations are the product of a regularly convened high-level ministerial forum, whereas the ATS functions predominantly at a technocratic level. The Arctic Council has adop-ted multiple declarations and texts addressing aspects of the Anthropocene; in the ATS, such documentary outcomes are rare. Even when ministerial or other high-level statements are adopted by the Antarctic Treaty parties (such as the 2009 Washington Ministerial Meeting Declaration on the Fiftieth Anniversary of the Antarctic Treaty, or the 2016 Santiago Declaration on the Twenty-fifth Anni-versary of the Environmental Protocol), they tend to be far more conservative documents than their Arctic equivalents.

The bureaucratisation of the ATS is one of the reasons for its success and resi-lience. In addition, one of the strengths of the ATS is often said to be its deeply embedded legal character, which stands in contrast to the more informal Arctic governance arrangements. However, legal codification can have drawbacks, as it can create cumbersome arrangements and 'path dependency'. Young (2017, p. 42) argues that, in the Anthropocene, there is a 'premium on the creation and opera-tion of international environmental and resource regimes that are effective in tur-bulent times and capable of adapting nimbly or agilely to rapidly changing conditions'. Environmental regimes, he contends, will only be effective if they can not only anticipate 'state changes in both biophysical and socioeconomic systems', but also 'take steps to avoid passing tipping points' (p. 42). When measured against these attributes, the ATS has clear deficiencies.

With its clear legal foundations and cooperative systems for Antarctic governance, the ATS has functioned well in maintaining a dialogue between the Antarctic Treaty parties, improving coordination and defusing tensions. This has served to protect the compromise over sovereignty enshrined in Article 4 of the Antarctic Treaty. Yet, there is an understandable anxiety about the stability of the regime, if the global profile of Antarctic governance were to be lifted significantly. Nonetheless, as a counterpoint, it should be noted that significant turning points in Antarctic governance have left the ATS in a stronger position; the debate over mining which led to the adoption of the Environmental Protocol being the most obvious example. Moreover, the ATS is now more robust than it has ever been, with a growing and active membership now comprising twenty-nine Consultative and twenty-four non-Consultative parties. Importantly, among this expanded membership are states that were critical of the Antarctic regime. For instance, Malaysia, which acceded to the Antarctic Treaty in 2011, had led efforts to place the 'Question of Antarctica' on the UN General Assembly Agenda from 1983, and had questioned the legitimacy of the Antarctic Treaty.[9] Given these developments, the ATS is now in a better position to engage with issues of global significance. Moreover, such engagement may serve to reduce the perception that the ATS has engendered an unhelpful degree of Antarctic exceptionalism.[10]

The problem of 'regime fit' means that the ATS itself cannot address the main threats Antarctica faces in the Anthropocene, from rising temperatures to changing ocean circulation. Addressing the damaging consequences of changes to the global carbon cycle remains the primary responsibility of the global climate regime, the UNFCCC. However, there are opportunities for strengthening the UNFCCC 'regime complex' (Keohane & Victor 2011) in which the ATS and the Arctic Council have influence that is proportionate to the impacts of climate change on the polar regions and the significance of the polar regions for the global climate system. A 'regime complex' (Raustiala & Victor 2004) refers to a governance arrangement in which multiple treaty or other management systems are connected in a formal or informal way to address a common issue or problem, or are applied to a defined geographical region (Young 2016, p. 218). If the UNFCCC is viewed in these more expansive terms, then there are opportunities for involvement and input from other treaty regimes, such as the ATS, that have a clear interest in the governance of global climate change.

While the ATS has been effective in enabling the detection and anticipation of state-change in Antarctica, principally by rendering Antarctica a *terra clima*, it has been mostly ineffective in contributing to global efforts to prevent this change. This was highlighted by the 2017 SCAR Science Lecture at ATCM 40 in Beijing, presented by New Zealand palaeoclimatolgist Tim Naish. Since 2003, the ATCM has featured lectures on Antarctic science arranged by SCAR, and Naish's 2017 lecture assessed the implications of the 2015 Paris Agreement for Antarctica. The 2017 SCAR Science Lecture was a wake-up call at ATCM 40, prompting several questions and comments which indicated that many parties are fully aware that major change in Antarctica is underway. There were also moments of honesty; for example, when an American diplomat described the crack in the Larsen C Ice Shelf as 'terrifying' during a special session of ATCM 40 arranged by China to showcase its Antarctic interests.

In his 2017 SCAR lecture, Naish highlighted the clear links between the Paris Agreement temperature goals and corresponding impacts on the physical Antarctic environment. He explained that the threshold for loss of Antarctica's stabilising ice shelves may be the Paris target of two degrees Celsius of global warming. If the world exceeds this, we will commit Planet Earth to multi-metre sea-level rise that may be irreversible for millennia (Naish 2017, n.p.). He argued that the Antarctic was now at a critical moment, and that there was a limited time in which to prevent unstoppable change. Naish noted that, despite the significant impacts of climate change on 'Antarctic activities and operations … the ATS does not have a coherent voice in the UNFCCC' (n.p.). To address this deficiency, he made several suggestions for greater practical engagement between the ATS and UNFCCC.

The UNFCCC system has not effectively internalised polar imperatives in its decision-making and goal-setting. Indeed, neither the UNFCCC nor the Paris Agreement refers to the polar regions, despite their central importance to the global climate system. Other regional collectives and groupings have been far more visible in the climate regime complex, such as small and low-lying developing states. These are specifically referenced in the UNFCCC,[11] and they have some influence upon climate negotiations. The polar regions face challenges that are different from this grouping of states and, notably, are not exposed to the existential threats which confront some low-lying small island states. Moreover, in the Antarctic context climate impacts do not involve direct effects on human wellbeing and livelihoods. In the Arctic, by contrast, climate change is affecting many indigenous groups, and indigenous people's organisations have been vocal advocates for greater action to address climate change within the Arctic Council (Young 2016, p. 213).

There are several possibilities, some small, some large, for greater ATS engagement with climate governance. Making enhanced Antarctic visibility in the UNFCCC a feasible proposition is the overlap between Antarctic Treaty Consultative Parties (ATCPs) and major emitters with the greatest influence within the climate regime complex. Of the world's ten largest carbon emitters (the People's Republic of China, the United States, India, Russia, Japan, Germany, Iran, Saudi Arabia, South Korea and Canada), eight are parties to the Antarctic Treaty, and seven of these are ATCPs.[12]

At the more modest end of the scale of policy responses, the ATCM could promote measures aligned with the Paris Agreement's two degrees Celsius temperature goal, such as committing to reduce to zero the carbon footprint of Antarctic operations. The 2010 ATME Climate Report dealt specifically with this issue, recommending that the parties 'promote reduction of the carbon footprint of activities in Antarctica' and recognising 'the importance of emission cuts in Antarctica and their symbolic value in the global context' (Secretariat of the Antarctic Treaty 2010, n.p.).[13]

More substantively, the ATCM could, through an appropriate resolution, articulate the ways in which the implementation of the ATS could help to achieve the Paris Agreement goals. For instance, the ban on mining in Antarctica could be highlighted, not only for its importance to Antarctic environmental protection, but also for its contribution to preventing dangerous climate change by keeping fossil fuels in the ground and staying within the two degrees Celsius carbon budget. This could have bipolar resonance, in light of estimates that all of the Arctic's hydrocarbon resources are unburnable under the Paris Agreement's temperature goals (McGlade & Ekins 2015).

More importantly still, the ATCM and CEP could specifically identify the implications of the Paris Agreement objectives for Antarctica, through resolutions affirming the importance of retaining Antarctica's ice shelves and sheets, and noting the CO_2 concentrations and temperature tipping points at which they will begin an inexorable collapse. This could build upon the example of ATCM Resolution XV-14 (1989) and Resolution 6 (2015), by adding specific guardrails and benchmarks now made possible by the extensive Antarctic scientific research undertaken since the 1990s. While the specification of such boundaries would have no normative force in and of itself, it could make a meaningful contribution to global climate governance by increasing the profile of Antarctica, and polar system changes generally, within the UNFCCC. This would complement and provide policy focus for the work of the Intergovernmental Panel on Climate Change (IPCC), which recently decided to commission a 'Special Report on the Ocean and the Cryosphere in a Changing Climate', to be finalised by 2019.

As part of a strategy to raise the visibility of Anthropocene Antarctica in the climate regime complex, there is value in revisiting proposals for more structured engagement by the ATS with UNFCCC negotiations, such as Australia's proposal at ATCM 35 in 2012 for Antarctic Treaty parties to 'pursue enhanced engagement in international discussions on climate change' (Commonwealth of Australia 2012). At ATCM 35, Australia submitted a working paper which detailed several options for ATS engagement with UNFCCC negotiations. These included registering the ATS as an observer organisation to the UNFCCC or a subsidiary body, issuing a joint statement on Antarctic issues to the UNFCCC COP and engaging in the work of the UNFCCC's Subsidiary Body for Scientific and Technological Advice as a partner organisation.

There are dozens of observers to the UNFCCC; yet, the ATS, which represents the largest component of the Earth's cryosphere, has no status or voice within global climate talks. As the Australian working paper explained, 'Pursuing closer engagement with the UNFCCC would be consistent with the provisions of the Antarctic Treaty, and with the practice of establishing effective working relationships with other international organisations where necessary to advance the protection and management of the Antarctic region' (Commonwealth of Australia 2012).

Australia's proposal received a lukewarm reception at the ATCM, with the meeting report indicating that, while most parties believed there was value in making the UNFCCC aware of Antarctic climate science, several parties 'raised concerns about the merits and costs of registering the ATS as an observer at the

UNFCCC, the merits of tasking the Secretariat with a policy liaison role, and the challenges of negotiating a statement by all Consultative parties' (ATCM 2012, p. 277). This response seemed to confirm the Antarctic Treaty Consultative Meeting as a debating chamber, while the real decisions of moment in the Anthropocene are being made elsewhere. However, given the renewed global attention on climate issues following the successful adoption of the Paris Agreement in 2015, the time is now ripe for the recommendations in this proposal to be revisited as part of a strategy to develop a meaningful ATS response to the Anthropocene.

Conclusions

The current preoccupations of the ATS lie with the control and mitigation of regional and local environmental impacts. Activities in Antarctica itself (including stationing, expeditions, research, fishing, tourism and potential mining) could, even if completely unregulated, occasion only a fraction of the damage being felt from climate change, ocean acidification and other global forces in the Anthropocene. It is almost as if the hyper-environmental focus of regional Antarctic governance is a 'coping strategy'; a fixation being pursued to compensate for the global failure to address the much larger and more urgent threats to Antarctica.

In its legal design and practical operation, the ATS has tended to remain aloof from and indifferent to global affairs. There are exceptions at certain defined moments, most notably at its inception and during the minerals debate in the 1980s, when global concerns could not be ignored. There is a widely held view that it is not possible, desirable or even necessary for the ATS to march into the territory of climate mitigation, or indeed to address other global environmental questions, such as the current negotiations on a new instrument under the 1982 UN Convention on the Law of the Sea to protect biodiversity beyond national jurisdiction (Van der Watt 2017, p. 587). However, the Anthropocene is an era in which, of necessity, the Antarctic will again become a focus of global environmental interest. The essential question for the Antarctic Treaty parties is whether this will sharpen sufficiently for a policy response only after it is too late to do anything about the global forces transforming Antarctica, or whether the ATS can make a vital contribution in the next decade, as the future of Antarctica is being determined.

This chapter argues for much closer integration of polar priorities within the operation of the UNFCCC and the Paris Agreement where the global carbon budget is determined. As part of that process, the ATS has a clear interest in ensuring that the timeline for addressing climate change takes the impacts on the Antarctic environment into account. The ATS is in the central position to advance this, and the process can begin in a relatively uncontroversial way through the policy-relevant science that forms the primary currency of discourse in the Antarctic system. An appropriate starting point would be the identification of polar-relevant climate governance targets; for example, regional temperature changes at which the loss of Antarctic ice sheets will become irreversible, or a pH range for a fully functioning Southern Ocean marine ecosystem. The successful achievement of the goals of the ATS and climate regimes is inextricably linked, yet this interdependency is not currently reflected in governance arrangements.

Acknowledgments

This chapter is a modified version of an earlier article published as Stephens, T 2018, 'The Antarctic Treaty System and the Anthropocene', *The Polar Journal*, vol. 8, no. 1, pp. 29–43, and is reprinted by permission of the publisher, Taylor & Francis Ltd, <http://www.tandfonline.com>.

Disclosure statement

The author attended the Fortieth Antarctic Treaty Consultative Meeting in Beijing, 22 May to 1 June 2017, as an academic observer with the Australian Government delegation.

Funding

The work was supported by the Australian Research Council Future Fellowship Scheme.

Notes

1 See also Aitken et al. 2016.
2 See also Lorimer 2015 and Hamilton 2016.
3 See Shaw et al. 2014, n.p., who note that the 'apparent protection status of [Antarctica] reflects management intent, not management outcome'.
4 CCAMLR Conservation Measure 24–04 (2016).
5 CCAMLR Conservation Measure 91–05 (2016).
6 See also ATCM Recommendation VIII-12 (1975) and Resolution 2 (1995), urging governments to prevent the disposal of nuclear waste in the Antarctic Treaty Area. It should be noted that the ATS does not expressly prohibit the use of nuclear reactors, such as the one used at McMurdo in the 1960s and 1970s.
7 See generally Stephens 2017, p. 31.
8 ATCM Decision 8 (2009). But cf. CCAMLR Resolution 30/XXVIII (2009) which asked the CCAMLR Chairman to write to the UNFCCC to convey CCAMLR's view that 'an effective global response by the UNFCCC is urgently needed to address the challenge of climate change'.
9 See further Saul & Stephens 2015.
10 On 'Antarctic exceptionalism', see Hemmings 2014.
11 UNFCCC, Preamble, and Article 4(8).
12 Greenhouse gas data drawn from the World Resources Institute 2014.
13 ATME Climate Change Report, Recommendations 4 and 5.

References

Aitken, ARA, Roberts, JL, van Ommen, TD, Young, DA, Golledge, NR, Greenbaum, JS, Blankenship, DD & Siegert, MJ 2016, 'Repeated large-scale retreat and advance of Totten Glacier indicated by inland bed erosion', *Nature*, no. 533, pp. 385–389.

Antarctic Treaty Consultative Meeting (ATCM) 2012, *Final Report of the Thirty-fifth Antarctic Treaty Consultative Meeting*, Secretariat of the Antarctic Treaty, Buenos Aires.

Antarctic Treaty Consultative Meeting (ATCM) 2015, *Final Report of the Thirty-eighth Antarctic Treaty Consultative Meeting*, Secretariat of the Antarctic Treaty, Buenos Aires.

Arctic Council 2013, Statement to UNFCCC COP19, viewed 13 March 2019, <http://www.arctic-council.org/index.php/en/our-work2/8-news-and-events/143-statement-copxix>.

Arctic Council 2017, Fairbanks Declaration, viewed 13 March 2019, <https://oaarchive.arctic-council.org/handle/11374/1910>.

Biermann, F 2014, *Earth system governance: world politics in the Anthropocene*, MIT Press, Cambridge, Mass.

Chakrabarty, D 2009, 'The climate of history: four theses', *Critical Inquiry*, vol. 35, pp. 197–222.

Chown, SL 2017, 'Antarctic environmental challenges in a global context' in K Dodds, AD Hemmings & P Roberts (eds), *Handbook on the politics of Antarctica*, Edward Elgar, Cheltenham, pp. 523–539.

Colgan, W, Machguth, H, MacFerrin, M, Colgan, JD, van As, D & MacGregor, JA 2016, 'The abandoned ice sheet base at Camp Century, Greenland, in a warming climate', *Geophysical Research Letters*, vol. 43, pp. 8091–8096.

Commission for the Conservation of Antarctic Marine Living Resources (CCAMLR) 2009, 'Resolution 30/XXVIII', viewed 13 March 2019, <https://www.ccamlr.org/en/resolution-30/xxviii-2009>.

Commission for the Conservation of Antarctic Marine Living Resources (CCAMLR) 2016, *Conservation Measures*, viewed 13 March 2019, <https://www.ccamlr.org/en/conservation-and-management/conservation-measures>.

Commonwealth of Australia 2012, 'ATCM interests in international climate change discussions – options for enhanced engagement', Working Paper 32, ATCM XXXV, CEP XV, Hobart.

DeConto, RM & Pollard, D 2016, 'Contribution of Antarctica to past and future sea-level rise', *Nature*, vol. 531, pp. 591–597.

Dryzek, JS, Norgaard, RB & Schlosberg, D 2013, *Climate challenged societies*, Oxford University Press, Oxford.

French, D & Scott, KN 2009, 'International legal implications of climate change for the polar regions', *Melbourne Journal of International Law*, vol. 10, pp. 631–655.

Gaffney, O & Steffen, W 2017, 'The Anthropocene equation', *The Anthropocene Review*, vol. 4, pp. 53–61.

Ganopolski, A, Winkelmann, R & Schellnhuber, HJ 2016, 'Critical insolation – CO2 relation for diagnosing past and future glacial inception', *Nature*, vol. 529, pp. 200–203.

Gill, N 2016, 'Rupture in Tasmania: solastalgia and the impact of the recent bushfires', *The Monthly*, April 2016, viewed 13 March 2019, <https://www.themonthly.com.au/issue/2016/april/1459429200/nicole-gill/rupture-tasmania>.

Hamilton, C 2015, 'Human destiny in the Anthropocene' in C Hamilton, C Bonneuil & F Gemenne (eds), *The Anthropocene and the global environmental crisis*, Routledge, Abingdon, pp. 32–43.

Hamilton, C 2016, 'The Anthropocene as rupture', *The Anthropocene Review*, vol. 3, pp. 93–106.

Hemmings, AD 2014, 'Re-justifying the Antarctic Treaty System for the 21st century: rights, expectations and global equity' in RC Powel & K Dodds (eds), *Polar geopolitics? Knowledges, resources and legal regimes*, Edward Elgar, Cheltenham, pp. 55–73.

Keohane, RO & Victor, DG 2011, 'The regime complex for climate change', *Perspectives on Politics*, vol. 9, no. 1, pp. 7–23.

Leane, E 2016, *South Pole: nature and culture*, Reaktion Books, London.

Lorimer, J 2015, *Wildlife in the Anthropocene: conservation after nature*, University of Minnesota Press, Minneapolis.

Matear, RJ, O'Kane, TJ, Risbey, JS & Chamberlain, M 2015, 'Sources of heterogeneous variability and trends in Antarctic sea-ice', *Nature Communications*, vol. 6, article no. 8656, doi:10.1038/ncomms9656.

McGlade, C & Ekins, P 2015, 'The geographical distribution of fossil fuels unused when limiting global warming to 2°C', *Nature*, vol. 517, pp. 187–190.

McNeill, JR & Engelke, P 2014, *The Great Acceleration: an environmental history of the Anthropocene since 1945*, MIT Press, Cambridge, Mass.

Naish, T 2017, 'ATCMXL SCAR Lecture: What does the United Nations Paris Climate Agreement mean for Antarctica?', viewed 13 March 2019, <https://scar.org/library/policy/antarctic-treaty/scar-lecture-presentations/3483-scar-lecture-2017/>.

Nash, S 2011, 'Persistent organic pollutants in Antarctica: current and future research priorities', *Journal of Environmental Monitoring*, vol. 13, pp. 497–504.

Nicholson, S & Jinnah, S 2016, 'Introduction: living on a New Earth', in S Nicholson & S Jinnah (eds), *New Earth politics: essays from the Anthropocene*, MIT Press, Cambridge, Mass., pp. 1–16.

Notz, D & Stroeve, J 2016, 'Observed Arctic sea-ice loss directly follows anthropogenic CO2 emissions', *Science*, vol. 354, no. 6313, pp. 747–750.

O'Reilly, J 2013, 'Antarctic climate futures: how terra incognita becomes terra clima', *The Polar Journal*, vol. 3, no, 2, pp. 384–389.

Österblom, H & Olsson, O 2017, 'CCAMLR: an ecosystem approach to the Southern Ocean in the Anthropocene' in K Dodds, AD Hemmings & P Roberts (eds), *Handbook on the politics of Antarctica*, Edward Elgar, Cheltenham, pp. 408–421.

Peden, J, Tain, T, Pertierra, LR & Tejedo, P 2016, 'Perceptions of the Antarctic wilderness: views from emerging adults in Spain and the United States', *Polar Record*, vol. 52, pp. 541–552.

Petrenko, VV, Smith, AM, Schaefer, H, Riedel, K, Brook, E, Baggenstos, D, Harth, C, Hua, Q, Buizert, C, Schilt, A, Fain, X, Mitchell, L, Bauska, T, Orsi, A, Weiss, RF & Severinghaus, JP 2017, 'Minimal geological methane emissions during the Younger Dryas-Preboreal abrupt warming event', *Nature*, vol. 548, pp. 443–446.

Raustiala, K & Victor, D 2004, 'The regime complex for plant genetic resources', *International Organization*, vol. 58, pp. 277–309.

Saul, B & Stephens, T 2015, 'Responsive Antarctic law-making in the Asian Century', *The Yearbook of Polar Law*, vol. 7, pp. 55–82.

Scientific Committee on Antarctic Research 2009, *Antarctic Climate Change and the Environment (ACCE) Report*, SCAR, Cambridge.

Secretariat of the Antarctic Treaty 1989, 'ATCM XV, Recommendation ATCM XV-14 (Paris, 1989)', viewed 13 March 2019, <https://www.ats.aq/devAS/info_mea sures_listitem.aspx?lang=e&id=170>.

Secretariat of the Antarctic Treaty 2010, *Antarctic Treaty Meeting of Experts (ATME) on Implications of Climate Change for Antarctic Management and Governance (ATME Climate Change Report)*, ATCM XXXIII, WP063 (Norway and United Kingdom), viewed 13 March 2019, <https://www.ats.aq/documents/ATME2010/fr/ATME2010_fr001_e.pdf>.

Secretariat of the Antarctic Treaty 2017, 'ATCM XL – CEP XX, Decision 1 (2017) - ATCM XL - CEP XX, Beijing', viewed 13 March 2019, <https://www.ats.aq/devAS/ats_meetings_meeting_measure.aspx?lang=e>.

Shaw, JD, Terauds, A, Riddle, MJ, Possingham, HP & Chown, SL 2014, 'Antarctica's Protected Areas are inadequate, unrepresentative, and at risk', *PLoS Biology*, vol. 12, no. 6, doi:10.1371/journal.pbio.1001888.

Solomon, S, Ivy, DJ, Kinnison, D, Mills, MJ & Neely, RR 2016, 'Emergence of healing in the Antarctic ozone layer', *Science*, vol. 353, pp. 269–274.

Stephens, T 2017, 'Reimagining international environmental law in the Anthropocene' in L Kotzé (ed.), *Environmental law and governance for the Anthropocene*, Hart Publishing, Oxford, pp. 31–54.

Tollefson, J 2016, 'Antarctic model raises prospect of unstoppable ice collapse', *Nature*, vol. 531, pp. 562–562.

Turner, J & Comiso, J 2017, 'Solve Antarctica's sea-ice puzzle', *Nature*, vol. 547, pp. 275–277.

UNESCO, UCS & UNEP 2016, *World Heritage and tourism in a changing climate*, UNEP, Nairobi.

National Research Council. 2014. *The Arctic in the Anthropocene: Emerging Research Questions*, The National Academies Press, Washington, DC.

Van Der Watt, L-M 2017, 'Contemporary environmental politics and discourse analysis in Antarctica' in K Dodds, AD Hemmings & P Roberts (eds), *Handbook on the politics of Antarctica*, Edward Elgar, Cheltenham, pp. 584–598.

Wadhams, P 2016, *A farewell to ice: a report from the Arctic*, Allen Lane, London.

Waller, CL, Griffiths, HJ, Waluda, CM, Thorpe, SE, Loaiza, I, Moreno, B, Pacherres, CO & Hughes, KA 2017, 'Microplastics in the Antarctic marine ecosystem: an emerging area of research', *Science of the Total Environment*, vol. 598, pp. 220–227.

Wolff, EW 2013, 'Ice sheets and the Anthropocene', *Geological Society London Special Publications*, vol. 395, pp. 255–263.

World Resources Institute 2014, 'CAIT– Country Greenhouse Gas Emissions Data', viewed 13 March 2019, <https://www.wri.org/resources/data-sets/cait-country-greenhouse-gas-emissions-data>.

Young, OR 2016, 'The shifting landscape of Arctic politics: implications for international cooperation', *The Polar Journal*, vol. 6, no. 2, pp. 209–223.

Young, OR 2017, *Governing complex systems: social capital for the Anthropocene*, MIT Press, Cambridge, Mass.

Zalasiewicz, J, Waters, C & Head, MJ 2017, 'Anthropocene: its stratigraphic basis', *Nature*, vol. 541, pp. 289–289.

Zalasiewicz, J, Waters, CN, Williams, M, Barnosky, AD, Cearreta, A, Crutzen, P, Ellis, E, Ellis, MA, Fairchild, IJ, Grinevald, J, Haff, PK, Hajdas, I, Leinfelder, R, McNeill, J, Poirier, C, Richter, D, Steffen, W, Summerhayes, C, Syvitski, JPM, Vidas, D, Wagreich, M, Wing, SL, Wolfe, AP & Zhisheng, A 2015, 'When did the Anthropocene begin? A mid-twentieth century boundary level is stratigraphically optimal', *Quaternary International*, vol. 383, pp. 196–203.

3   Subglacial nationalisms

Alan D. Hemmings

Introduction

Antarctic geopolitics is a tussle with various materialities (see e.g. Antonello 2017, p. 167). Amongst these are geographical space, involving positions around territorial sovereignty, national 'presence' and the historical contingency of areas of Antarctic operation; and scientific research, which serves multiple purposes, from objective interest in regional to global contexts to instrumental need (perhaps nowhere more evident than in relation to anthropogenic climate change), Antarctic institutional glue, economic enabler, surrogate for national interest and laundry for nationalism. The concept of the Anthropocene,[1] an Earth-era in which human agency has become the prime mover, arises primarily from concern about anthropogenic climate change.[2] The cascade of consequential global system transformations is, whatever its wider applicability and intellectual merits, a concept almost perfectly tailored for contemporary Antarctic geopolitics. It is grounded in the data and analysis of the physical sciences, which grants it status and standing in the context of the privileged position science occupies in the Antarctic politico-legal system; and its leading scientific theorists themselves construe the Anthropocene as having clear implications for governance and social values globally (see e.g. Steffen et al. 2018). This fits the norms of Antarctic science as a benign contemporary 'civilising mission' – or voice for 'planetary stewardship' (see Castree 2014, pp. 245–248) – in a region blissfully clear of human occupants. As such, the Anthropocene provides a further social or public-interest justification for the ongoing conduct of science in Antarctica, one of the key geographic regions for research into anthropogenic climate change. Critically perhaps, as a concept still under development, with divergent views of its rigour, utility and implications, it does not come freighted with clear directions for policy-makers (or at least there is not yet consensus amongst policy-makers that it does).[3] The concept does not presently inconvenience, far less challenge, any other interests in Antarctica. If the Anthropocene can be seen by some as requiring collective or universalist responses, it can be seen by others as perfectly compatible with nationalist framings of state interest and engagement. In relation to Antarctica as much as anywhere else, the Anthropocene has, to use Yadvinder Malhi's words, become 'a scientific and cultural zeitgeist, a charismatic mega-category emerging from and encapsulating elements of the spirit of our age' (2017, p. 2).

This chapter casts the Anthropocene as essentially just a contemporary background to what is, frankly, business-as-usual in the Antarctic – the instrumental use of Antarctic activities to advance essentially *political* objectives. It examines the interface of nationalism with a particular subset of 'subglacial research', a term of art which I have employed to cover a range of research foci, including penetration of lakes and other features deep below Antarctica's ice cap (Siegert et al. 2013), drilling through that ice cap to recover ice cores for the elucidation of past climates (Summerhayes 2015),[4] and *sui generis* projects such as 'IceCube', where arrays of sensors are embedded deep in the ice sheet in order to detect neutrinos from outer space as they pass through the earth (University of Wisconsin-Madison n.d.).

Subglacial research might, at first glance, seem an unlikely object for capture by Antarctic nationalisms,[5] having (in public terms at least) an arcane focus, and modalities literally and conceptually deeper and more remote than those ordinarily seen to attract nationalist attention worldwide (see generally Breuilly 2013). Its material focus transcends regional particularity, makes sense only in broader contexts and purposes, requires intellectually and logistically complex and very expensive 'big science',[6] and thus seems inherently to call for internationalist approaches and joint projects. But science 'as currency' within the Antarctic Treaty System,[7] the totemic value of big science as a marker for wider national interests and prestige, the public appropriability and status signification of 'million-year' ice cores or subglacial lake penetration, both individually and collectively, facilitate an Antarctic nationalism (see generally Hemmings et al. 2015). Interestingly, this arises without explicit rejection of the established norms of internationalism in Antarctic scientific research. Indeed, the nationalism is invariably disguised (if it is not entirely denied), and the projects for which it may be a significant driver are loudly marketed as contributions to such international public goods as scientific knowledge and the reinforcement of scientific cooperation under the auspices of the Antarctic Treaty System.[8] David Mason captures this perfectly when he writes: 'It is within this context [the Antarctic Treaty as a model of international co-operation and stewardship] that Australia, as both a claimant state and an active supporter of the Antarctic Treaty System, negotiates and balances its interests and frames its laws' (Mason 2018).

A critical leitmotif in Antarctic affairs is the *dual-purpose* activity. Whilst a fuller explication of the historic place and a comprehensive examination of the contemporary manifestations of dual-purpose activities in Antarctica and the Antarctic politico-legal regime cannot be attempted in this chapter,[9] 'dual-purpose' in relation to subglacial research is, I believe, pertinent to the purposes of the 'Depths and Surfaces' theme of the conference at which a preliminary version of this chapter was presented and its situation in the Anthropocene. The organisers – the Humanities and Social Sciences Expert Group of the Scientific Committee on Antarctic Research – enjoined us, *inter alia*, to 'work on many levels', to draw 'attention to the three-dimensional nature of the ice sheet at a time when reports of increased glacial melt are appearing almost daily, and ice core sampling is helping us to understand long-term climate patterns', and (particularly relevant) 'taken metaphorically, … go beyond surface readings of a place which, in the past, has often been uncritically labelled a "last wilderness" or "continent for science" and considered outside the purview of the [Humanities and Social Sciences] disciplines' (Humanities and Social Sciences Expert Group 2017, p. 2). As Juan Francisco Salazar notes, an ice core 'becomes not an inert object but something more active: a lively subject communicating across time and space with a variety of audiences' (2018, p. 38). Developing a theme earlier raised by Klaus Dodds in a paper on 'Awkward Antarctic nationalism' (2017) which considered, *inter alia*, ice cores, Salazar goes on to remark that this 'raises to the fore awkward encounters for claimant states in terms of how the planetary history of the earth becomes embedded in nationalist narratives and practices' (2018, p. 38).

My proposition is that subglacial research, whatever else it is, is now also a mechanism co-opted in the pursuit of nationalist agendas in Antarctica. Whilst I believe that the nationalism angle can be seen to be operative across the range of subglacial research – and the fact of this generally being 'big science' is perhaps just part of the reason why this is so – space limitations necessarily impose constraints. Accordingly, here I only address the ice core projects and I do so through the prism of one state: Australia.

I do this for four main reasons:

(1) There is a greater number of ice core drilling projects (at least twelve – see Figure 3.1 below) than other subglacial research projects.[10] Hence they offer some basis for comparison, and opportunities to examine different modes of operation.

(2) These projects concern what must be recognised as the most compelling contemporary research need on our planet – understanding climate change – and thus most clearly encapsulate what one might term the 'public good' rationale for Antarctic science, so widely employed since the inception of the Antarctic Treaty System, and within which science is granted such a prominent role (Elzinga 2017a).[11]

(3) There is a fascinating case history emerging right now involving Australia – the state making the largest territorial claim in Antarctica.[12] As Antarctica's most assertive and nationalist claimant (see the argument in Dodds & Hemmings 2009) – which at the same time is a strong player in the Antarctic Treaty System[13] – Australia has one of the highest Antarctic expenditures,[14] and has significant capabilities and profile in relation to the conduct of Antarctic science (including climate change research).[15]

(4) At least part of the Australian case history has played out in the public domain. Whilst Australian Cabinet and other national high-policy documents may not presently be publicly available, a substantial number of Ministerial press releases and government agency policy documents are publicly accessible. Furthermore, there has been a high level of media coverage (at least by Antarctic standards) of Australian and other states' ice core projects. So, one is in a position not only to track what is being done in the field so to speak, but also to see how this has been framed in the public domain.

Australia is also the state in which I resided for fifteen years, before moving back to New Zealand in 2018. Accordingly, I am aware that some Australian Antarctic officials – whether current or former officials and science managers – can be sensitive (and sometimes arguably a little thin-skinned) about critiques of Australian Antarctic policy. Whilst I am not apologetic about my critique, it might be helpful for me to be clear about the scope of that critique. I am *not* contesting the case for subglacial research in general or ice core drilling projects in particular; neither am I contesting the particular technologies judged essential for the realisation of such projects. Elucidation of the processes of anthropogenic climate change is essential, and it is entirely reasonable that Australia contribute to this. It is of course a sad truth that in Australia, as in most other states, our research efforts in relation to climate change are not remotely matched by government policy responses. My interest is a deconstruction of a contributory factor in the preferred mode of Australian engagement, which I see as being indicative of a pervasive Australian Antarctic nationalism.[16]

Antarctic nationalisms

'Nationalism' is a term with diverse uses. These range from the structural and perhaps now anodyne underpinning of the emergence of nation states, through various historic emancipatory projects involving peoples, ethnic identities and geographical spaces, to 'banal' (Billig 1995), 'everyday' (Benwell 2014), 'awkward' (Dodds 2017), and even 'dim-witted' (Williams 2017) manifestations. Nationalism segues into various other concepts: the various 'isms', including 'colonialism', 'imperialism' and 'orientalism'. Attempting a comprehensive single definition of 'nationalism' is thus complex and accordingly any particular definition is likely to be contested.

As used here, it is as a deeply and widely embedded *ideology* that sees one's particular national group (for simplicity's sake I will construe this as captured by the term 'State', although agency resides in both civil society and the state *sensu stricto*) as not only having particular *interests* – there is nothing peculiar about that, every state has what it generally calls its 'national interest' – but *special* and *superior* interests in relation to any other state. Nationalism is many things, but it is invariably a claim to *privileged rights* in relation to something. And, generally, that something involves a geographical space. The supposed privileged rights are invariably predicated on a narrative which melds various presumed historic, social, ethnic/racial and religious standings.

Nationalism is conceded by most scholars to have been evident in Antarctica at various times *in the past*: most obviously during the Heroic Era; in the period immediately before the Second World War; during the period of tripartite brinkmanship in the Antarctic Peninsula in the 1940s and early 1950s – the period that gave rise to the framing of 'The Antarctic Problem' (Christie 1951) – and so on, right up to the adoption of the Antarctic Treaty in 1959. At which point, it ceased to be polite to notice it. In recent years, and, in my own estimation, particularly since the end of the Cold War, this has changed. Nationalism is now conceded by many to be behind what *others* (but not *we*) are *really* up to in Antarctica today. Continuing a project conducted with a number of colleagues internationally,[17] the focus of this chapter is nationalism in (and in relation to) Antarctica *today*. It is something that *we* are doing, today, even if *others* are also doing it.

Notwithstanding the exotic fantasy that some sort of substantive and realisable prospects sit behind any of the seven territorial claims to parts of Antarctica (or the basis of claim of the two semi-claimants of the United States and Russia), a critical difference between 'Antarctic Nationalism' and (most) 'Nationalism Elsewhere' is (ironically) its very *extra-territoriality*. Antarctica is not anybody's genuine 'home' as that term is usually understood. So, notwithstanding some historic engagement there, and the identification with place that this may well have engendered, the drivers of Antarctic nationalism largely seem to involve *aspiration* for standing, prestige and resources (as these are variously construed), rather than primarily a locally grounded (and locally generated) deeply historical and sanctified past. There is, after all, no permanent residential population in any normal sense, and so there is no community in the place itself which can trace lineage, tradition and cultural identity to the place.[18] No state or people has substantial engagement with Antarctica dating before the early twentieth century, and most only rather later.

Antarctic nationalism is particularly evident amongst the seven territorial claimants; indeed, the fact of being a claimant seems often the primary driver of that nationalism, whosoever within that state is articulating it. However, even the earliest of the territorial claims dates only to 1908; the last to the 1940s. Territorial positions have now been internationally 'frozen' (through Article IV of the 1959 Antarctic Treaty) and Antarctica instead has now operated as a *de facto* condominium for sixty years, compared with the fifty-one years between 1908 and 1959 – during which period both the United States and the Soviet Union actively repudiated the claims. The outcome of any testing of the status of these generally unrecognised claims in legal tribunals (most likely the International Court of Justice) post-Antarctic Treaty[19] is surely uncertain, given these realities and the fast-changing global order. But, beyond these legal matters, the *moral* basis for the claimants' continued assertion of territorial rights seems thin, to put it mildly (see Hemmings 2009; Mancilla 2018).[20] So, claimant Antarctic nationalism may have a touch of 'whistling Dixie' about it. To be clear, Antarctic nationalism is certainly not confined to claimants, but it is arguable that some (not all) Antarctic nationalism demonstrated in non-claimant states is a *reaction* to claimant positions. Thus, claimancy is a critical driver of contemporary Antarctic nationalism generally.

In an earlier paper (Hemmings et al. 2015), eleven ways in which Antarctic nationalisms might arise were identified. These are listed below (Box 3.1), slightly rephrased and reordered, with examples of possible actors indicated at the right. In practice, actual nationalistic stances are likely to involve multiple bases, coalescing in various manners. The last two (10 National pride in, and mobilisation through, national Antarctic programmes; and 11 Infrastructure and logistics arrangements) have particular salience for this chapter.

Box 3.1 Bases for Antarctic nationalism

1   Formally declared claims to territorial sovereignty – the seven claimants and two semi-claimants
2   Proximity of Antarctica to a state's metropolitan territory – Argentina, Australia, Chile, New Zealand, South Africa (and the relic colonial territories in the subantarctic of European states – France, Norway and the United Kingdom)
3   Historic and institutional associations with Antarctica – any Antarctic-active states
4   Social and cultural associations – state and civil society media, arts
5   Regional or global hegemonic inclinations – British Empire (historic), USA, USSR (now Russian Federation), China, etc.
6   Alleged need for resources – G20 states and others
7   Denial or constraint of access by strategic competitors or opponents – the 'West' in relation to China, and contrariwise
8   Contested uses or practices in Antarctica – Japan vs Australia, New Zealand, etc., around whaling; the West or Anglosphere vs Russia and China in relation to Antarctic Marine Protected Areas (MPAs)
9   Carry-over from intense antipathies outside Antarctica – Argentina/UK (Falklands/Malvinas), Russia/Ukraine (Crimea, Eastern Ukraine)
10   National pride in, and mobilisation through, national Antarctic programmes
11   Infrastructure and logistics arrangements

(Modified after Hemmings et al. 2015)

Ice core drill sites

There are a dozen ice core drill sites across the continent operated by various states, or groups of states (Figure 3.1).

Six of these sites are located in that part of East Antarctica that Australia claims as the 'Australian Antarctic Territory' (AAT):[21]

1   Dome A (77° E) – Chinese led
2   Vostok (106° E) – Russian led
3   Law Dome (112° E) – Australian led
4   Dome C (123° E) – European led
5   Taylor Dome (158° E) – US led
6   Talos Dome (159° E) – European led

Whilst there is a high level of international collaboration and participation (at the level of research scientists) across all these drill sites, only one project (Law Dome) has been led by Australia.

Figure 3.1 Ice core drill sites

Elsewhere across the continent, Dome Fuji is a Japanese project; all the other sites are collaborative European or US-led. Aant Elzinga (2017b, p. 217) has remarked that 'Owing to the steep costs, only a few countries have their own major ice coring projects: Australia, Japan and the United States'. He may have overlooked Russia, but the general point remains valid and raises the interesting question of what is it that would drive Australia – a 'middle-power' in its own estimation[22] – or even Russia, to incur the sort of financial and other costs that are taken up only by the world's major economic powers, rather than seeking to participate in a joint project?

Ice cores and nationalism in the 'Australian Antarctic Territory'

National Interests

Australia has been consistent, clear and transparent in relation to its 'national interests' in Antarctica. Other groups, from think tanks (e.g. Bergin & Haward 2006; Fogarty 2011; Bergin 2016) to senior serving officers in the Defence Force (Forster 2016), have also considered what Australia's national interests are, or should be. In this respect, Australia has been far clearer than many other Antarctic-active states, whether claimant or not. The United States, for example, makes its Presidential Directive on Arctic policy publicly available, but not the directive concerning the Antarctic (Hemmings 2012, p. 82). For many states, it is unclear whether they even have a formal Antarctic policy position informing their actual practice.

The Australian Government identifies its national interests in Antarctica under seven heads.[23]

'Australia's national interests in Antarctica are to:

1   maintain Antarctica's freedom from strategic and/or political confrontation
2   preserve our sovereignty over the Australian Antarctic Territory, including our sovereign rights over adjacent offshore areas
3   support a strong and effective Antarctic Treaty system
4   conduct world-class scientific research consistent with national priorities
5   protect the Antarctic environment, having regard to its special qualities and effects on our region
6   be informed about and able to influence developments in a region geographically proximate to Australia
7   foster economic opportunities arising from Antarctica and the Southern Ocean, consistent with our Antarctic Treaty system obligations, including the ban on mining and oil drilling.'

(Australian Antarctic Division [AAD] n.d. a)

So, Australia has a clear and unsurprising menu of national interests in Antarctica. There is nothing very remarkable about this, and most other Antarctic-active states would have something similar, even if not explicitly codified in a document. Other claimants (and the semi-claimants) will have a comparable national interest in maintaining their sovereignty positions. Non-claimants will obviously not have this interest, although maintaining their position in relation to claims by others will likely be its analogue.[24]

The Australian Antarctic Strategy and 20 Year Action Plan

A sequence of announcements and events commenced in 2013 around the project of developing an *Australian Antarctic Strategy and 20 Year Action Plan*. It will be necessary to set out some of the key stages as this unfolded.

The strategy and plan arose through a process formally initiated by the Liberal/National Government under then Prime Minister Tony Abbott, who was sworn into office on 18 September 2013. The new government announced its decision to proceed with this Antarctic strategic plan in Hobart just over a month later, on 28 October, when Environment Minister Greg Hunt revealed not only the terms of reference for the project, but also that it would be led by Tony Press, a former Director of the Australian Antarctic Division (Hunt 2013).[25] His report was to be delivered to the Australian Government by July 2014.

Consistent with the title of the Minister's media release – 'Tasmania will benefit from 20 Year Australian Antarctic Strategic Plan' – in it he referred to Hobart as a research 'hub' and noted that:

The Government has already committed significant new investment to support Tasmania's Antarctic role. This includes $38 million for the extension of Hobart International Airport, $24 million to establish a new Centre for Antarctic and Southern Ocean Research, and $25 million for the Antarctic Climate and Ecosystems Cooperative Research Centres.

(Hunt 2013)

The Commonwealth Government's interests in Tasmania are plainly many and various, and in recent years shoring-up electoral interests in that economically weak state have been amongst these.

Critically, for the focus of this chapter, half of the Minister's media release related to Australian participation in a

major multi-nation project which will reveal Antarctic climate records. The Australian Antarctic Division's Dr Mark Curran will lead a team of 15 partner organizations from Australia, China, Denmark, France, Germany and the United States to drill a 2000 to 3000-year ice core at Aurora Basin in east Antarctica. Over six weeks, beginning December 2013, 24 scientists will drill a 400 metre-long ice core at the remote site, 550 kms from Australia's Casey station … The traverse, to be joined by the Australian Antarctic Division's head of Climate Processes and Change, Dr Tas van Ommen, will be managed by a 14-strong team of colleagues from the French station Dumont d'Urville.

(Hunt 2013)

The report by Tony Press – identified as 'Head Inquirer', but with no other 'inquirers' identified – (Press 2014), duly appeared in July 2014. The Press Report underpinned the *Australian Antarctic Strategy and 20 Year Action Plan* which the government released in 2016 (Commonwealth of Australia 2016), with a joint media release from the Prime Minister, Foreign Minister and Minister for the Environment, entitled 'A new era of Antarctic engagement' (Turnbull, Bishop & Hunt 2016). The joint media release noted, *inter alia*, that 'A deep-field overland science traverse capability and mobile research station infrastructure will provide essential support for critical Antarctic science' (p. 1).

The Prime Minister, in his Foreword to the *Antarctic Strategy and 20 Year Action Plan*, wrote that 'The Government is delivering a new era of Australian Antarctic endeavour. It is timely to reaffirm our national Antarctic interests, and to set out a plan to protect and promote them' (Commonwealth of Australia 2016, p. 1). The *Antarctic Strategy and 20 Year Action Plan* lists the seven Australian national interests already referred to above.[26] But it operationalises these through what it identifies as 'Key actions the Government will deliver' (p. 3):

- A new world-class research and resupply Antarctic icebreaker.
- New and stable funding to support an active Australian Antarctic programme.
- Establish Australia's position of science leadership in Antarctica through:

 a developing modern and flexible infrastructure, including

- restoring traverse capabilities and establishing mobile stations in the Antarctic interior
- further scoping options for expanded aviation capabilities to establish a year-round aviation capability between Hobart and Antarctica
- progressing options for more efficient and flexible use of existing research stations

 b a revitalised science programme, including

- coordinated and effective funding of Antarctic science
- opportunities for public-private partnerships to conduct new and iconic scientific research endeavours

 c greater collaboration and resource-sharing with other nations active in East Antarctica.

- Strengthen the Antarctic Treaty system and our influence in it, by building and maintaining strong and effective relationships with other Antarctic Treaty nations through our international engagement.
- Build Tasmania's status as the premier East Antarctic Gateway for science and operations, including through:

 a streamlined Government regulatory and approval processes to facilitate increased use of Hobart as an Antarctic Gateway port

 b agreeing priority proposals with industry to enhance Tasmania's status as an Antarctic Gateway, including expanded infrastructure in Hobart for the new icebreaker

 c a major review on building research infrastructure in Hobart to establish Australia as the world's leader in krill research.

(Commonwealth of Australia 2016, p. 3)

The *Antarctic Strategy and 20 Year Action Plan* clearly addresses a broad swathe of Australian national interests in Antarctica. Before returning to the direct connections between it and Australian ice coring on the polar plateau, it is necessary to consider quite why million-year ice cores are of interest.

The scientific case for Million-Year Ice Cores

The 'Million-Year Ice Core' has a substantive significance in addition to the public-relations and political significance that will be examined below. The 2011 Antarctic Science Advisory Committee (ASAC)[27] explains it thus:

There is currently insufficient knowledge of the climate system to allow models to reproduce past glacial cycles from orbital inputs alone. Very old ice cores could provide the answer to a key puzzle: why glacial interglacial cycles occurred with 41 thousand year periodicity prior to one million years ago and gradually shifted to a strong 100 thousand year periodicity for most of the last million years. Currently the longest ice core records extend to around 800 thousand years, but if ice over a million years can be drilled, this will help assess the role that carbon dioxide feedback played in this shift. This is relevant to understanding whether the current human-driven rise in carbon dioxide beyond past interglacial levels will drive the earth system into another semi-stable state that does not involve glacial advances and retreats. If it is established that sufficiently old ice exists in Antarctica to study this transition, it is almost certain to be within the AAT.

(Antarctic Science Advisory Committee 2011, pp. 34–35)

So, there is a scientific case for seeking to acquire such cores; East Antarctica (a large part of which is claimed by Australia as the AAT) is plainly a place where this is possible; and therefore Australian engagement in such a project is entirely reasonable. Interestingly, the 'Million-Year Ice Core' label, which has proven so popular for politicians and journalists, seems to have been promoted by the scientific community itself from quite early on. Certainly, in the Australian context, this label appears in the media release from the conference of the International Partnerships in Ice Core Sciences (IPICS) held in Hobart in 2016 (Antarctic Climate and Ecosystems Cooperative Research Centre 2016). This media release is clearly the source for the contemporaneous 'Million-year-old ice' and 'holy grail' news reports (Carlyon 2016; Slezak 2016), which in turn defined the language to be used in most subsequent Australian media reports, including those of ABC journalist Jane Norman, below (2016a; 2016b).[28] This language is now also routinely used by Antarctic scientists and science managers, including the (then) Director of the Australian Antarctic Division (Gales 2017).

Joining the threads: leadership – traverses – ice cores

There are plainly a number of threads that one could follow from the *Antarctic Strategy and 20 Year Action Plan*. But the one that interests me here is the one tracing from that hardy perennial of *leadership* ('science leadership')[29] so beloved of politicians, and a supposed route to this through 'modern and flexible infrastructure', a component of which is apparently 'restoring traverse capabilities and establishing mobile stations in the Antarctic interior' (Commonwealth of Australia 2016, p. 3). When combined with the commitment to achieving a 'Million Year Ice Core' (p. 14),[30] this is the thread that takes us from a worthy enough and unremarkable (if perhaps banal) political statement to substantial expenditure and the enabling of an Australian ice core drilling project near Dome C, East Antarctica, in what Australia calls the 'Australian Antarctic Territory'. Indeed, this thread is evident in the more detailed planning sections of the *Antarctic Strategy and 20 Year Action Plan* in 'Year Two' (p. 24),[31] 'Year Five' (p. 25),[32] and 'Years 10–20' (p. 26),[33] following the release of the Action Plan. Whilst the planning and subsequent interpretation phases are both described in a context of working with 'international partners', this is not mentioned in the context of the middle phase ('Year Five') – i.e. when actually acquiring the million-year ice core. That phase appears (so far as the Action Plan is concerned) to be construed as an Australia-only phase.

What is going on here – below the surface both literally and metaphorically? The deep answer to this, in my judgement, requires us to understand that multiple considerations are in play. First, there is the historic Australian concern – notwithstanding its declaratory confidence in the robustness and continuing validity of its claim to forty-two percent of the continent – that its claim to the 'Australian Antarctic Territory' may be a bit thin away from the immediate environs of its coastal stations. This traces back particularly to the legal analysis of Gillian Triggs, who concluded that whilst Australia may indeed have valid title to areas effectively occupied by it around Davis, Casey and Mawson stations, and thereby adjacent continental shelf (Triggs 1986, p. 322), there is 'little evidence to support Australian sovereignty over the vast hinterland of its claimed sector beyond exploratory expeditions and the extension of legislation. It is thus doubtful whether Australia can support its claim to sovereignty over such territory' (p. 323). This is plainly a disquieting legal opinion for a claimant. There have been two sorts of response in the decades since Triggs' opinion: seeking alternative legal assessments that may offer more encouraging news (a process that is ongoing); and seeking to change the facts on the ground – i.e. trying to demonstrate effective occupation in polar terms by increasing deep field activities.

Of course, other states' activities have also increased in East Antarctica since the publication of Triggs' analysis in 1986. This increased activity has included these 'other' states' operations deep inland in the claimed sector — particularly, the Russians (continuing earlier Soviet practice) and Chinese, but also fellow-claimant France (the Franco-Italian Concordia Station is in the claimed 'Australian Antarctic Territory', not in the French-claimed Terre Adélie). China is the greatest concern for Australia in relation to its own assessment of its position with regard to the claimed 'Australian Antarctic Territory'. One sees this expressed across policy documents, parliamentary considerations of Australia in Antarctica,[34] and in the media. Media commentary is generally sourced to staff at think tanks, such as the Australian Strategic Policy Institute, former officials and sometimes Antarctic scientists (who would not dream of commenting on another *scientific* discipline).[35] This China concern is of course neither confined to Antarctica, nor to Australia. But, in the Antarctic context, the sentiment is well represented in the 2014 Press Report:

> Australian leadership in Antarctica and the Southern Ocean is eroding. As Australia's logistic and scientific capabilities stagnate through historical erosion of funding and the aging of its assets, other countries are ramping up their investments in Antarctic science, logistics and infrastructure. In Australia's area of direct interest China, Republic of Korea and India have expanded their Antarctic investments in recent years. China in particular is expanding its efforts in East Antarctica and the Southern Ocean, including in the Australian Antarctic Territory. Since 1996 China has expanded its

Zhongshan station near Australia's Davis station, and built Kunlun, an inland station at Dome A, the highest place in the Australian Antarctic Territory. China has built a new summer station between Zhongshan and Kunlun; a new icebreaker; and a fifth Antarctic station is going to be built in the Ross Sea to the East of the Australian Antarctic Territory.

(Press 2014, p. 21)

India and South Korea are in the frame, one feels, without any great concern merely to create the appearance of an equality of treatment with China – whose activities alone are itemised.

A third consideration is the talismanic significance of big-science and associated big-logistics such as *traverses*. Think, *inter alia*, of the profile of the Fuchs/Hillary Trans Antarctic Expedition, the Soviet Vostok traverses and the US Pole Route in Antarctic prestige and, presumptively, influence terms – and in domestic political and national self-esteem terms. Alongside this is the very high public profile and political significance (whether positive or negative) of climate change research – perhaps particularly for governments disinclined or feeling unable to take substantive policy action – and the genuine and enduring scientific interest in doing such research of particular individuals, communities and agencies.

If one were seeking a project that would enable one to demonstrate activity in the back-blocks of the claimed 'Australian Antarctic Territory', fit well within the science-first norms of the Antarctic Treaty System, compete with the Chinese and anybody else who might have the temerity to operate in 'our' territory, and display Australian technical prowess, a deep field million-year ice core would seem to tick the boxes.

Traverse – ice core – race!

Whatever the substantive scientific justifications, the publicity value of a 'Million-Year Ice Core' has been considerable – hence the appearance of the term. Deployed across the international polar community, this artefact – an imagined as much as actual destination, like the poles of an earlier age – has facilitated the emergence of a sense of great purpose, an international project and (perhaps paradoxically in strictly scientific terms, where it matters not who wins) a *race* between competing national teams.[36]

A consequence of a *race* framing is that it undercuts joint projects. One needs one's own team; one cannot race unless there is a *competitor*. And that appears to mean that whilst the Australian drilling project – like the European one – identifies Dome C as a good site, one does not combine with the Europeans or even base oneself at what a reasonable-minded observer might think is the 'conveniently located' Concordia Station.[37] Could that decision have anything to do with the fact that, although the Franco-Italian Concordia Station is in the right place, it is in the territory claimed by Australia? Further, what does *not* appear to have been asked – or at least not asked or explained in the public domain – is why science needs *two* ice core projects at Dome C? If there is a case, it has not been made.

If one does not support the project out of Concordia, short of building one's own new station (a project likely to cost many hundreds of millions of dollars), one has to use a *field camp* (which may in fact be as sophisticated as a main station was fifty years ago!), thereby demonstrating one's capacity, commitment, 'effective occupation' credentials and thus, as a claimant, one's enduring *claimantness* (if I may be pardoned the neologism). It offers, *inter alia*, a corrective to the criticism from some Australian think tanks and media that a spending/capacity/credibility gap between Australia and China in particular has emerged in 'our' Antarctic territory (see e.g. Bergin 2014). Inevitably this all turns out to be quite expensive. The Australian Government has found A$45 million to fund it.

> The Coalition Government committed $45 million dollars earlier this year to develop an overland traverse capability to pursue the million year ice core as outlined in the *Australian Antarctic Strategy and 20 Year Action Plan*. The funding covers ice core drilling equipment and mobile station infrastructure and support, so Australia can take a lead role in the project.
>
> (Frydenberg 2016)[38]

The territorial imperative revisited

The odd thing is that, territorially at least, nobody's actions – Australia's, China's or any other Party's – have significance whilst the Antarctic Treaty is in force, according to Paragraph 2 of Article IV of the 1959 Antarctic Treaty – which is generally seen as the keystone article of the Treaty.

> 2. No acts or activities taking place while the present Treaty is in force shall constitute a basis for asserting, supporting or denying a claim to territorial sovereignty in Antarctica or create any rights of sovereignty in Antarctica. No new claim, or enlargement of an existing claim, to territorial sovereignty in Antarctica shall be asserted while the present Treaty is in force.
>
> (Secretariat of the Antarctic Treaty n.d., p. 22)

All claims, bases of claims or potential claims were, in effect, suspended as of the entry into force of the Treaty in 1961 and nothing which occurs while the Treaty is in force will affect the pre-existing position of all of the interested parties – both the claimants and the non-claimants (Rothwell & Hemmings 2018, p. xxi). This is in marked contrast to the situation in the 1980s in the era of the 'Question of Antarctica' at the UN General Assembly. Then the friction was between insider ATS members and a critical external community of states (the G77 in the 1980s), the latter not bound to the undertakings within the Antarctic Treaty System. The contemporary contestation is between what are often – and not accurately – cast as 'traditional' Antarctic Treaty Consultative Parties (ATCPs) and 'new' ATCPs. Critically, this means that, notwithstanding the attempt by some to suggest a sort of ranking or caste system within the ATCPs (1. 'original signatories'; 2. other long term ATCPs; and 3. states from the global south and elsewhere which became ATCPs 'only' in the 1980s), *all* of the notional protagonists are parties to the ATS and bound by its obligations. These obligations include those of Article IV of the Antarctic Treaty. So, the question of whether efforts seemingly *contra* Paragraph 2 of Article IV are legitimate against the contingency of third parties is essentially moot. The main players in the Antarctic 'Great Game' are already parties to the ATS and legally equal in relation to its rights and duties.

This seems to provide a nice example of the role of Antarctic science as 'Symbolic Political Capital' (Elzinga 2017a, pp. 105–106). Whilst Elzinga deployed this assessment generically across Antarctic science, in my view it has often been most evident in Antarctic claimant states. An interesting facet of the Australian Antarctic discourse over the past decade has been the increasing frequency with which the instrumental value of Antarctic *science* for Australian Antarctic *policy* aspirations has been publicly declared. Sometimes this may be little more than opportunistic special pleading – as when Antarctic scientists argue the utility of their research (often the imperative of being funded to do the research!) in terms that they hope will cut through to scientifically ignorant purse holders. Sometimes it seems grounded on not very much at all beyond optimism (see e.g. McCallum 2017). But elsewhere it reflects the fact that Australia's Antarctic science is actually formally structured to support the national policy position. The Australian Antarctic Division (a division of the Commonwealth Department of the Environment and Energy) has a brief that is not confined to the conduct of science.

> The Division is responsible for the, 'Advancement of Australia's strategic, scientific, environmental and economic interests in the Antarctic by protecting, administering and researching the region'.
>
> (See 'About us', AAD b.)

One sees intimations of the instrumental function in section 3 (Australia's Administration of the Australian Antarctic Territory) of the 2014 Press Report, where it is suggested that it is 'appropriate for Australia's operational and scientific presence' that:

> Australia should ensure that it rebuilds its deep field traverse capability to support multi-national logistic and science efforts in the Antarctic, and *demonstrate Australian leadership*.
>
> … Australia should be a significant player or leader in large, multi-national research efforts in the Australian Antarctic Territory. *Because of the important status that science is afforded in the Antarctic Treaty*, Australia's ability to participate in and lead key science programs in the Australian Antarctic Territory is not only a signal of Australia's commitment to the Treaty, but is also a demonstration of Australia's engagement in the region.
>
> (Press 2014, pp. 25–26, emphasis added)

So, rather clearly, Australian Antarctic science is viewed as a mechanism that can and should be deployed to further Australia's political interests in Antarctica – central to which is its continuing claim to what it calls the 'Australian Antarctic Territory'.

Conclusion: reflections on the contemporary Antarctic

What is this all about? Why do we see a state operating like this in Antarctica in the second decade of the twenty-first century? When one deconstructs activity-sets such as this subglacial research, we seem to end up a very long way from the traditional conception of Antarctica as the continent for science, wherein it was assumed that the operative activity (science) was a nice, neutral, common-interest sort of activity; and that in Antarctica we were in the happy situation that 'science' had replaced the familiar contestation between states that we see elsewhere. Of course, this was always nonsense and scholars such as Elzinga have carefully documented why this is so. This exploration of the linkage between subglacial research and nationalism in Antarctica is therefore, amongst other things, the addition of a further case study of Antarctica's essentially geopolitical dimension.

'Subglacial' relates not only to the physical location of the ostensible activity on the continent of Antarctica but also to the deep layering of the drivers of that activity and the pathway from objective scientific purpose through valid instrumental imperatives, such as understanding (and, critically, responding to) climate change, and the interactions (whether synergistic or antagonistic) between research, governmental and non-governmental interests, to subjective (and not necessarily positive) nationalist purposes.

Whilst subglacial research has substantive scientific, environmental and, as a consequence, (hopefully) a positive public policy value, it also has a nationalist dimension. States are prepared, in Antarctica, to incur higher costs in relation to such 'big science' research in order to *also* pursue prestige, 'leadership' and pro-file;[39] and (in their estimation at least) to underpin their geopolitical interests. Notwithstanding the second paragraph of the Antarctic Treaty's Article IV, state practice suggests that some of them appear to believe that activities 'on the ground' in Antarctica (so to speak) may in fact have probative value (at least in geopolitical terms) in relation to their future standing in Antarctica regarding territory – although they may be disinclined to claim any formal legal basis for such a position. What one sees with subglacial research appears to be a further demonstration of the dual-function/use conundrum in Antarctic affairs. A sort of plausible deniability attaches to activities which can be laundered as purely scientific – and are indeed entirely consistent with international cooperation and collaborative science – but which in fact *also* have as part of their deep purpose the furtherance of a political project, the realisation of which might in fact unpick the delicate Antarctic *modus vivendi*.

And what of that 'Million-Year Ice Core', that 'holy grail' that Australia pursued? A 2.7 million-year ice core was recovered by a US team in another part of Antarctica in 2015, and reported in 2017 (Voosen 2017). This ice was 'stratigraphically disturbed sections of ice up to 2.7 million years old' from blue ice in the Allan Hills in East Antarctica (Kehrl et al. 2018). Of course, that ice can be argued to be not quite the same as a continuous ice core covering the entire period back one million years, and indeed a site has now been examined at the Allan Hills that offers hope of a continuous record going back through a million years (Kehrl et al. 2018). But, whilst this may indeed be a significant consideration for the glaciologists, it is likely too fine a distinction to cut much ice in the public relations game. In terms of winning *that* race, the Australian project had, it appears, 'lost' before it got out of the starting blocks.

Notes

1 For a review of the concept, and the diversity of disciplinary engagements with it, see Malhi 2017.
2 Although it may also be seen to capture other facets of human activity which are not necessarily direct consequences of climate change – in Antarctica, the effects of marine harvesting within the region and pollution by chemicals or micro-plastics arriving from elsewhere. See Stephens in this volume, pp. 18–19.
3 For some thoughts on the implications of the Anthropocene for the development of international law, see Hey 2016; and in relation to the operation of the Antarctic politico-legal system, see Stephens in this volume.
4 See Summerhayes 2015, particularly ch. 13 'Solving the ice age mystery: the ice core tale'.
5 Nationalisms plural – see 'Antarctic nationalisms'.
6 'Big science' is used here for projects of scale and complexity. They generally entail interdisciplinarity and involve large numbers of scientists and support personnel, often from multiple agencies and states, and may extend over several years or be effectively ongoing. Elzinga identifies an important platform for the emergence of big science in Antarctica (although he does not use the term explicitly) when '[f] rom the late 1980s Antarctic science was gradually integrated into what was later called Integrated Earth Systems Science' (2017a, pp. 115–117). For an example of a recent climate-change related big science project, see the BBC report on UK and US research cooperation at Thwaites Glacier (Amos 2018).
7 The term memorably coined to describe the fact that Antarctic science had emerged as a main means of exerting 'practical influence in Antarctic affairs' (Herr & Hall 1989, p. 13).
8 See, *inter alia*, the Preamble and Articles II and III of the Antarctic Treaty (Secretariat of the Antarctic Treaty n.d., pp. 21–22).
9 Perhaps most evident in relation to interpretative practice concerning 'measures of a military nature' *sensu* Article 1 of the Antarctic Treaty (Hemmings 1990; Dodds & Hemmings 2008, p. 175), including alleged dual-use satellite earth stations operated in Antarctica by Norway, China and others (Wormald 2011; Darby 2014), but also arguably in relation to 'environmental management' (Hemmings 2013).
10 Despite there being over 400 subglacial lakes, only three exploration projects have so far been undertaken (see Siegert et al. 2016).
11 The politics of science, and science in polar politics, have been examined in detail over several decades by Aant Elzinga. See e.g. Elzinga 2017a.
12 Australia currently claims forty-two percent of the continent.
13 Various indices might be adduced; e.g. diplomatic papers at Antarctic Treaty Consultative Meetings (Dudeney & Walton 2012); Australia providing two of the five Chairs of the Committee for Environmental Protection since its first meeting in

1998; hosting the Secretariat of the Commission for the Conservation of Antarctic Marine Living Resources (CCAMLR), etc.

14 Anne-Marie Brady places Australia at US$150 million, behind only the United States (see Brady 2012, p. 12).

15 Australia ranks highly in the number of scientific papers produced. See Dastidar & Ramachandran 2008; Gray & Hughes 2016.

16 See, *inter alia*, Dodds & Hemmings 2009; Leane et al. 2016; Dodds 2017. Australian and other Antarctic nationalisms are also recurrent themes in Dodds et al. 2017.

17 Involving, *inter alia*, Sanjay Chaturvedi, Klaus Dodds, Elizabeth Leane, Daniela Liggett and Juan Francisco Salazar.

18 This does not of course preclude nationalist advocacy from within the relatively large community of individuals who have transiently, if sometimes repeatedly, resided at Antarctic facilities, nor the development there of what O'Reilly and Salazar (2017) have termed 'non-sovereign national enclaves'.

19 Positions are held in abeyance for the duration of the Antarctic Treaty under its Article IV.

20 For a wider critique of Antarctic territorial claims, see Hemmings 2012; Scott 2015.

21 The AAT claim comprises two sectors, one between 45° E and 136° E and the other between 142° E and 160° E.

22 This is a theme fashionable in Australian international relations discourse in recent years. See e.g. Cotton & Ravenhill 2011.

23 These interests are listed at a number of government agency sites, including that of the Australian Antarctic Division (Australian Antarctic Division n.d a).

24 For example, non-claimants Germany, India, Japan and the Netherlands (in addition to semi-claimants, the Russian Federation and the United States) lodged objections to Australia's Antarctic Extended Continental Shelf submission (Hemmings & Stephens 2010, p. 318).

25 One might also reflect that the choice of a former senior science manager to lead the project signalled that it would – formally, if not in reality – be predicated on technical rather than politico-legal considerations.

26 For an analysis of *The Antarctic Strategy and 20 Year Action Plan* as a whole, see McGee & Smith 2017.

27 In the interest of transparency, it should be noted that I was a member of ASAC in 2011.

28 There is a broader question, which cannot be examined in this chapter, about the role of the media – at the extreme, becoming complicit – in the mobilisation of Antarctic nationalism. See, for example, the coverage of the Norwegian Prime Minister's visits to Antarctic by the newspaper *Aftenposten* (Alnæs 2017).

29 Commonwealth of Australia 2016, p. 19. 'Leadership' forms a key component of 'Advancing our national Antarctic interests', where three of the four ways in which Australia is to pursue its interests in Antarctica are 'leadership' focused: 'leadership and influence in Antarctica', 'Leadership and excellence in Antarctic science', 'Leadership in environmental stewardship in Antarctica'. The fourth is 'Develop economic, educational and collaborative opportunities'.

30 Commonwealth of Australia 2016, p. 14, but developed further as an imperative elsewhere.

31 Commonwealth of Australia 2016, p. 24: 'building an overland traverse capability, with associated ice core drilling and mobile station infrastructure and research support, to enable planning with international partners to retrieve a million-year ice core'.

32 Commonwealth of Australia 2016, p. 25: 'Established an overland traverse and mobile inland station, and commenced involvement in a major scientific research undertaking to retrieve a million-year old ice core'.

33 Commonwealth of Australia 2016, p. 26: 'Work with international partners to interpret the findings from the completed project to retrieve a million-year ice core'.

34 In 2014, for example, the Foreign Affairs, Defence and Trade References Committee reported that

> In particular, the growing profile of China as an Antarctic actor was mentioned frequently to the committee. China joined the Antarctic Treaty in 1983, but its engagement was relatively modest until this century. In the last ten years China has significantly increased its investment in the Antarctic region, including more than doubling spending on Antarctic science and logistics, and building new bases on the continent itself, including in the Australian Antarctic Territory.
>
> (2014, para. 2.12)

35 See the detailed consideration in Dodds & Hemmings 2009.
36 See e.g. the media framing and language attributed to senior scientists in Denholm 2016.
37 The difficulties around sharing station facilities are explored in Hemmings 2011.
38 See also the stories by an ABC political correspondent (not a science journalist) who was on a media visit to Antarctica, and whose story was filed on the day of the Minister's media release: Norman 2016a; 2016b.
39 Note that the 'Key actions' in the Antarctic Strategy and Action Plan include 'opportunities … to conduct new and *iconic* scientific research endeavours' (Commonwealth of Australia 2016, p. 3, emphasis added).

Acknowledgements

A shorter and earlier version of this chapter was presented at the Depths and Surfaces conference in Hobart in July 2017. I thank Elle Leane and her colleagues for the invitation to attend, and the participants for helpful feedback. My thanks also to the two anonymous reviewers for their comments on the first draft of this chapter, and to Jeff McGee for helpful comments on the revised draft. I thank Wolfgang Rack for his assistance in identifying the ice core drill sites for Figure 3.1. Over the past several years I have worked with others on various aspects of Antarctic nationalism – most notably Sanjay Chaturvedi, Klaus Dodds, Elle Leane, Daniela Liggett and Juan Francisco Salazar – to each of whom I extend my thanks for insights, inspiration and good company. Naturally all positions, arguments and errors in this chapter are solely the responsibility of the present author.

References

Alnæs, J 2017, 'The Prime Minister of Antarctica: entrenching territory through journalistic travelogues', *Studies in Travel Writing*, vol. 21, no. 4, pp. 403–418.

Amos, J 2018, 'Thwaites Glacier: biggest ever Antarctic field campaign', *BBC News*, 30 April 2018, viewed 29 March 2019, <https://www.bbc.com/news/science-environment-43936372>.

Antarctic Climate and Ecosystems Cooperative Research Centre 2016, 'In search of Earth's oldest ice: Hobart plays host to international ice core scientists', International Partnerships in Ice Core Sciences Second Open Science Conference, media release, viewed 29 March 2019, <http://www.antarctica.gov.au/news/2016/in-search-of-earths-oldest-ice>.

Antarctic Science Advisory Committee 2011, *Australian Antarctic Science Strategic Plan 2011–12 to 2020–21*, viewed 29 March 2019, <http://www.antarctica.gov.au/__da ta/assets/pdf_file/0019/27307/AASSP_final-published-version_Apr-2011.pdf>.

Antonello, A 2017, 'Life, ice and ocean: contemporary Antarctic spaces' in K Dodds, AD Hemmings & P Roberts (eds), *Handbook on the politics of Antarctica*, Edward Elgar, Cheltenham, pp. 167–182.

Australian Antarctic Division n.d. a, 'Australia's national interests in Antarctica', viewed 29 March 2019, <http://www.antarctica.gov.au/about-us/antarctic-strategy-and-a ction-plan/australian-antarctic-strategy/australias-national-interests-in-antarctica>.

Australian Antarctic Division n.d. b, 'About us', viewed 29 March 2019, <http://www.antarctica.gov.au/about-us>.

Benwell, M 2014, 'Connecting southern frontiers: Argentina, the South West Atlantic and "Argentine Antarctic Territory"' in RC Powell & K Dodds (eds), *Polar geopolitics? Knowledges, resources and legal regimes*, Edward Elgar, Cheltenham, pp. 201–214.

Bergin, A 2014, 'Funds needed to back Australian strategy for Antarctic cold rush', *Australian*, 13 October 2014.

Bergin, A 2016, 'Australia needs to strengthen its strategic interests in Antarctica', *Sydney Morning Herald*, 28 April 2016, viewed 29 March 2019, <http://www.smh.com.au/comment/australia-needs-to-strengthen-its-strategic-interests-in-antarctica -20160428-gogzic.html>.

Bergin, A & Haward, M 2006, *Frozen assets: securing Australia's Antarctic future*, Strategic Insights 34, Australian Strategic Policy Institute, Canberra.

Billig, M 1995, *Banal nationalism*, Sage, London.

Brady, A-M 2012, 'Polar stakes: China's polar activities as a benchmark for intentions', *China Brief*, vol. 12, no. 14, pp. 11–15.

Breuilly, J (ed.) 2013, *The Oxford handbook of the history of nationalism*, Oxford University Press, Oxford.

Carlyon, P 2016, 'Million-year-old ice the "holy grail" of climate research in Antarctica', *ABC News*, 7 March 2016, viewed 29 March 2019, <http://www.abc.net.au/news/2016-03-07/search-for-world27s-oldest-ice-core-continues-in-antarctica/7225734>.

Castree, N 2014, 'The Anthropocene and the environmental humanities: extending the conversation', *Environmental Humanities*, vol. 5, pp. 233–260.

Christie, EWH 1951, *The Antarctic problem*, George Allen & Unwin, London.

Commonwealth of Australia 2016, *Australian Antarctic Strategy and 20 Year Action Plan*, Australian Government, Canberra, viewed 29 March 2019, <http://www.antarctica.gov.au/__data/assets/pdf_file/0008/180827/20YearStrategy_final.pdf>.

Cotton, J & Ravenhill, J (eds) 2011, *Middle power dreaming: Australia in world affairs 2006–2010*, Oxford University Press in association with the Australian Institute of International Affairs, Melbourne.

Darby, A 2014, 'China's Antarctic satellite base plans sparks concerns', *Sydney Morning Herald*, 12 November 2014, viewed 29 March 2019, <http://www.smh.com.au/world/chinas-antarctica-satellite-base-plans-spark-concerns-20141112-11l3wx.html>.

Dastidar, PG & Ramachandran, S 2008, 'Intellectual structure of Antarctic science: a 25-years analysis', *Scientometrics*, vol. 77, no. 3, pp. 389–414.

Denholm, M 2016, 'Antarctica research race on for million-year-old ice extraction', *Australian*, 12 December 2016, viewed 29 March 2019, <http://www.theaustralian.com.au/national-affairs/climate/antarctica-research-race-on-for-millionyearold-ice-ex traction/news-story/a0018ecc98077b6d22376e64551bb6a9>.

Dodds, K 2017, '"Awkward Antarctic nationalism": bodies, ice cores and gateways in and beyond Australian Antarctic Territory/East Antarctica', *Polar Record*, vol. 53, no. 1, pp. 16–30.

Dodds, K & Hemmings, AD 2008, 'The United States 2002 Unified Command Plan: Antarctica and the areas of responsibility of military commanders', *Polar Record*, vol. 44, no. 2, pp. 173–177.

Dodds, K & Hemmings, AD 2009, 'Frontier vigilantism? Australia and contemporary representations of Australian Antarctic Territory', *Australian Journal of Politics and History*, vol. 55, no. 4, pp. 513–529.

Dodds, K, Hemmings, AD & Roberts, P (eds) 2017, *Handbook on the politics of Antarctica*, Edward Elgar, Cheltenham.

Dudeney, JR & Walton, DWH 2012, 'Leadership in politics and science within the Antarctic Treaty', *Polar Research*, vol. 31, doi:10.3402/polar.v31i0.11075.

Elzinga, A 2017a, 'The continent for science' in K Dodds, AD Hemmings & P Roberts (eds), *Handbook on the politics of Antarctica*, Edward Elgar, Cheltenham, pp. 103–124.

Elzinga, A 2017b, 'Polar ice cores: climate change messengers' in BB Vincent, S Loeve, A Nordmann & A Schwarz (eds), *Research objects in their technological setting*, Routledge, Abingdon, pp. 215–231.

Fogarty, E 2011, 'Antarctica: assessing and protecting Australia's national interests', *Policy Brief*, Lowy Institute for International Policy, viewed 29 March 2019, <https://www.files.ethz.ch/isn/132284/Fogarty,%20Antarctica_web.pdf>.

Foreign Affairs, Defence and Trade References Committee 2014, *Australia's future activities and responsibilities in the Southern Ocean and Antarctic waters*, Commonwealth of Australia, Canberra.

Forster, M 2016, 'Australia's national interests in the Antarctic region: what is important?', *Indo-Pacific Strategic Papers*, Centre for Defence and Strategic Studies, Canberra, viewed 29 March 2019, <http://www.defence.gov.au/ADC/Publications/IndoPac/Forster_IPSP_Final.pdf>.

Frydenberg, J 2016, 'Australia on track to find the world's oldest Antarctic ice', media release, 12 December 2016, viewed 29 March 2019, <https://www.environment.gov.au/minister/frydenberg/media-releases/pubs/mr20161212a.pdf>.

Gales, N 2017, 'Celebrating 70 years of the Australian Antarctic Program', *DFAT Blog*, 28 August 2017, viewed 29 March 2019, <https://blog.dfat.gov.au/2017/08/28/celebrating-70-years-of-the-australian-antarctic-program/>.

Gray, AD & Hughes, KA 2016, 'Demonstration of "substantial research activity" to acquire consultative status under the Antarctic Treaty', *Polar Research*, vol. 35, doi:10.3402/polar.v35.34061.

Hemmings, AD 1990, 'Is Antarctica demilitarised?' in RA Herr, HR Hall & MG Haward (eds), *Antarctica's future: continuity or change?*, Australian Institute of International Affairs, Hobart, pp. 225–241.

Hemmings, AD 2009, 'Beyond claims: towards a non-territorial Antarctic security prism for Australia and New Zealand', *The New Zealand Yearbook of International Law*, vol. 6, pp 77–91.

Hemmings, AD 2011, 'Why did we get an International Space Station before an International Antarctic Station?', *The Polar Journal*, vol. 1, no. 1, pp. 5–16.

Hemmings, AD 2012, 'Security beyond claims' in AD Hemmings, DR Rothwell & KN Scott (eds), *Antarctic security in the twenty-first century: legal and policy perspectives*, Routledge, London, pp. 70–94.

Hemmings, AD 2013, '"Environmental management" as diplomatic method: the advancement of strategic national interest in Antarctica' in D Liggett & AD Hemmings (eds), *Exploring Antarctic values*, Gateway Antarctica Special Publications Series 1301, Christchurch, pp. 70–89.

Hemmings, AD & Stephens, T 2010, 'The extended continental shelves of sub-Antarctic islands: implications for Antarctic governance', *Polar Record*, vol. 46, pp. 312–327.

Hemmings, AD, Chaturvedi, S, Leane, E, Liggett, D & Salazar, JF 2015, 'Nationalism in today's Antarctic', *The Yearbook of Polar Law*, vol. 7, pp. 531–555.

Herr, RA & Hall, HR 1989, 'Science as currency and the currency of science' in J Handmer (ed.), *Antarctica: policies and policy development*, Centre for Resource and Environmental Studies, Australian National University, Canberra, pp. 13–24.

Hey, E 2016, 'International law and the Anthropocene', *ESIL Reflections*, vol. 5, no. 10, pp. 1–7.

Humanities and Social Sciences Expert Group 2017, *Depths and surfaces: understanding the Antarctic region through the humanities and social sciences*, HASSEG conference 2017, 5–7 July, Hobart, viewed 29 March 2019, <http://antarctica-hasseg.com/wp-content/up loads/2017/05/Conference-BookletFINAL.pdf>.

Hunt, G 2013, 'Tasmania will benefit from 20 Year Australian Antarctic Strategic Plan', media release, 28 October 2013, viewed 29 March 2019, <http://www.environm ent.gov.au/minister/hunt/2013/pubs/mr20131028a.pdf>.

Kehrl, L, Conway, H, Holschuh, N, Campbell, S, Kurbatov, AV & Spaulding, E 2018, 'Evaluating the duration and continuity of potential climate records from the Allan Hills Blue Ice Area, East Antarctica', *Geophysical Research Letters*, vol. 45, pp. 4096–4104.

Leane, E, Winter, T & Salazar, JF 2016, 'Caught between nationalism and internationalism: replicating histories of Antarctica in Hobart', *International Journal of Heritage Studies*, vol. 22, no. 3, pp. 214–227.

Malhi, Y 2017, 'The concept of the Anthropocene', *Annual Review of Environment and Resources*, vol. 42, pp. 77–104.

Mancilla, A 2018, 'The moral limits of territorial claims in Antarctica', *Ethics & International Affairs*, vol. 32, no. 3, pp. 339–360.

Mason, D 2018, 'The rules-based order applying to Antarctica', *Australian Outlook*, 26 June 2018, viewed 27 March 2019, <https://www.internationalaffairs.org.au/australia noutlook/the-rules-based-order-applying-to-antarctica/>.

McCallum, A 2017, 'As China flexes its muscles in Antarctica, science is the best diplomatic tool on the frozen continent', *The Conversation*, 14 November 2017, viewed 29 March 2019, <http://theconversation.com/as-china-flexes-its-muscles-in-anta rctica-science-is-the-best-diplomatic-tool-on-the-frozen-continent-86059>.

McGee, J & Smith, D 2017, 'Framing Australian Antarctic policy: the 20-year Antarctic plan and beyond', *Australian Journal of Maritime and Ocean Affairs*, vol. 9, no. 1, pp. 25–41.

Norman, J 2016a, 'What is the "holy grail of climate science" and how will scientists find it?', *ABC News*, 12 December 2016, viewed 29 March 2019, <http://www.abc. net.au/news/2016-12-12/what-is-the-million-year-ice-core/8102552>.

Norman, J 2016b, 'Antarctica: Australian scientists prepare to unveil "holy grail of climate change" amid $45m funding boost', *ABC News*, 12 December 2016, viewed 29 March 2019, <http://www.abc.net.au/news/2016-12-12/scientists-prepare-to-un veil-holy-grail-of-climate-change/8110380>.

O'Reilly, J & Salazar, JF 2017, 'Inhabiting the Antarctic', *The Polar Journal*, vol. 7, no. 1, pp. 9–25.

Press, AJ 2014, *20 Year Australian Antarctic Strategic Plan*, viewed 29 March 2019, <http://www.antarctica.gov.au/__data/assets/pdf_file/0008/178595/20-Year-Plan_Press-Report.pdf>.

Rothwell, DR & Hemmings, AD 2018, 'Introduction: the context of international polar law' in DR Rothwell & AD Hemmings (eds), *International polar law*, Edward Elgar, Cheltenham, pp. xiii–xliv.

Salazar, JF 2018, 'Ice cores as temporal probes', *Journal of Contemporary Archaeology*, vol. 5, no. 1, pp. 32–43.

Scott, K 2015, 'Looking back to look forward: trends, challenges and scenarios for the Antarctic Treaty System in 2115' in P Kennedy (ed.), *The Arctic and Antarctica: differing currents of change*, New Zealand Institute of International Affairs, Wellington, pp. 95–107.

Secretariat of the Antarctic Treaty n.d., 'The Antarctic Treaty', 1 December 1959 (in force 23 June 1962), 402 UNTS 71, viewed 29 March 2019, <https://www.ats.aq/documents/keydocs/vol_1/vol1_2_AT_Antarctic_Treaty_e.pdf>.

Siegert, MJ, Kennicutt, MC & Bindschadler, RA (eds) 2013, *Antarctic subglacial aquatic environments*, Geophysical Monograph Series 152, American Geophysical Union, Washington DC.

Siegert, MJ, Priscu, JC, Alekhina, IA, Wadham, JL & Lyons, WB 2016, 'Antarctic sub-glacial lake exploration: first results and future plans', *Philosophical Transactions of the Royal Society A: Mathematical, Physical and Engineering Sciences*, vol. 374, doi:10.1098/rsta.2014.0466.

Slezak, M 2016, 'Climate scientists step up search for "holy grail" of million-year-old ice', *Guardian*, 9 March 2016, viewed 29 March 2019, <https://www.theguardian.com/world/2016/mar/09/climate-scientists-step-up-search-for-holy-grail-of-million-year-old-ice>.

Steffen, W, Rockström, J, Richardson, K, Lenton, TM, Folke, C, Liverman, D, Summerhayes, CP, Barnosky, AD, Cornell, SE, Crucifix, M, Donges, JF, Fetzer, I, Lade, SJ, Scheffer, M, Winkelmann, R & Schellnhuber, HJ 2018, 'Trajectories of the Earth System in the Anthropocene', *Proceedings of the National Academy of Sciences*, vol. 115, no. 33, pp. 8252–8259.

Summerhayes, CP 2015, *Earth's climate evolution*, Wiley-Blackwell, Chichester.

Triggs, GD 1986, *International law and Australian sovereignty in Antarctica*, Legal Books, Sydney.

Turnbull, M, Bishop, J & Hunt, G, 2016, 'A new era of Antarctic engagement', media release, 27 April 2016, viewed 29 March 2019, <http://www.environment.gov.au/minister/hunt/2016/pubs/mr20160427.pdf>.

University of Wisconsin-Madison n.d., *IceCube. South Pole Neutrino Observatory*, viewed 29 March 2019, <http://icecube.wisc.edu/>.

Voosen, P 2017, '2.7-million-year-old ice opens window on past', *Science*, vol. 357, no. 6352, pp. 630–631.

Williams, Z 2017, 'Homophobia is back – it's no accident that nationalism is too', *Guardian*, 3 July 2017, viewed 29 March 2019, <https://www.theguardian.com/commentisfree/2017/jul/02/homophobia-nationalism-china-us-difference>.

Wormald, B 2011, *The satellite war*, CreateSpace Independent Publishing Platform.

4 Frozen Eden lost?

Exploring discourses of geoengineering Antarctica

Jeffrey McGee

The history of human interaction with the Antarctic continent[1] is relatively short. It was first observed by the Russian explorer Bellingshausen during the early nineteenth century and, later that century, by various Southern Ocean whalers, sealers and maritime expeditioners. The early twentieth century saw the Heroic Era of intensive human exploration, including early mapping of the Antarctic land and ice sheets. In these years, humans reached the South Pole and South Magnetic Pole in the now famous (and at times tragic) expeditions of Scott, Shackleton, Amundsen, Mawson and others. More extensive human exploration of the Antarctic continent occurred during the early decades of the twentieth century, including extensive mapping of the ever-changing ice coastline. In the early 1950s, the human presence on the Antarctic continent entered a new phase of permanent human habitation, commencing with construction of the Mawson scientific research base by Australia in East Antarctica.

Seventy years later, the human presence on the Antarctic continent is still only modest but increasing. Permanent scientific bases on the Antarctic continent have increased to over fifty, comprising thirty-nine hectares of buildings and disturbing a further 520 hectares of land or ice (Brooks et al. 2019). Despite an increase in science and associated logistical personnel at research bases over winter, the human presence on the Antarctic continent during this period is only about 1,000, which increases to 4,000 in the summer months. However, the scale of Antarctic tourism has increased markedly over the last two decades, with over 50,000 cruise ship tourists now visiting the areas around West Antarctica during each Southern summer. But as Philpott (elsewhere in this volume) describes, at least in popular culture and understanding, there is still a continuing notion of separateness between the Antarctic cryosphere and the bulk of human activity on this planet.

It is highly questionable, however, whether this notion remains accurate. As discussed in the introduction to this volume, the concept of 'the Anthropocene' signifies the onset of a new geological era, in which human activity is identified as a major cause of biophysical change to the planet (Young 2017). This idea of human activity as a major cause of global biophysical change is well illustrated by the notion that there are 'planetary boundaries' of human activity (see Rockström et al. 2009 and Nash et al. 2017) that must be observed to maintain a 'safe operating space' for humanity. If the Earth is indeed in a new geological era defined by human activity as a key driver of biophysical change at a planetary scale, the traditional notion of separateness between humanity and the Antarctic continent needs to be reconsidered.

This task is heightened by the key messages from climate change science regarding the experienced warming of the planet from increased carbon dioxide concentration in the atmosphere and the outlook for significant further warming. Anthropogenic greenhouse gas emissions from human use of fossil fuels since the onset of the Industrial Revolution are now implicated in conditions that are giving rise to major biophysical change to the Antarctic continent. There is significant evidence that ice sheet deterioration in West Antarctica is associated with climate change (Antarctic Climate & Ecosystems Cooperative Research Centre [ACE CRC] 2017). The impact of human activity on planetary systems is therefore now manifest in the physical systems of the Antarctic continent. This will have ramifications for hundreds, if not thousands, of years.

This chapter highlights an important way in which climate change is further breaking down the sense of separateness between human activity and Antarctica. The social dynamic analysed here is the extension of human discourse of responding to climate change to include proposals for 'Antarctic geoengineering'. Dryzek (2006, p. 1) explains that 'discourse' refers to 'a shared set of concepts, categories, and ideas that provides its adherents with a framework for making sense of situations, embodying judgments, assumptions, capabilities, dispositions, and intentions. It provides basic terms for analysis, debates, agreements, and disagreements'. In this way, discourses define the limits of human thought, imagination and options for possible action in a given issue area. Following a 2009 report by the Royal Society of the United Kingdom, the term 'geoengineering' is widely understood to refer to proposals for 'deliberate large-scale manipulation of the planetary environment to counteract anthropogenic climate change' (The Royal Society 2009, p. 1). The report identifies two major categories of geoengineering proposals. First, solar radiation management (SRM) proposals, which seek to increase the Earth's albedo to reflect into space a small amount of incoming sunlight energy. Second, carbon dioxide removal (CDR) proposals, which seek to remove carbon dioxide from the atmosphere and sequester it in the land or oceans. As discussed below, the geoengineering discourse now extends beyond these two original categories to proposals for regional-scale human intervention in biophysical systems of the Antarctic cryosphere and Southern Ocean.

Proposals for geoengineering the Antarctic cryosphere are novel and in their very early days. Geoengineering is generally considered by its supporters to be an unfortunate, but necessary agenda for research, in the face of the climate crisis. It is generally not viewed as a replacement for mitigation of greenhouse gas emissions, but a policy supplement. Critics of geoengineering say that it is another example of human technological hubris which has little real prospect of success and many environmental and social risks (Thiele 2018). Geoengineering is likely to be particularly controversial if ever conducted in Antarctica, due to its image as a separate, environmentally pristine place.

The aim of this chapter is not to specifically advocate for or critique geoengineering. Instead, the chapter aims to show how geoengineering discourses are being applied to Antarctica and the way these discourses are already having an important impact in naturalising the idea (already evident in the Anthropocene) of Antarctica as a landscape that is influenced by humans on a large scale. There has been no attempt so far to survey the current geoengineering proposals that are directly relevant to Antarctica. The chapter fills this gap by outlining these proposals for Antarctic geoengineering and the way in which they evidence an important emerging discourse on the human response to climate change. The chapter therefore demonstrates that the Anthropocene is reshaping Antarctica in both a material and a discursive sense.

This chapter proceeds as follows. The first section explores links between the Antarctic cryosphere and Southern Ocean in the context of climate change. Here, we see representations of Antarctica as a scientific laboratory crucial to the understanding of climate science. We also see Antarctica as a repository of a vast amount of the world's fresh water and therefore a source of human danger from sea-level rise in a warming planet. The first section also provides an overview of various proposals for SRM and CDR geoengineering as part of the human response to climate change. The second section specifically examines key proposals for solar radiation management that might be used to protect the Antarctic cryosphere from the effects of climate change. These proposals range from global-scale interventions in the stratosphere to localised interventions that might be used at a smaller scale during periodic heat events. The third section explores proposals for sequestering carbon dioxide in nearby areas of the Southern Ocean. The fourth section looks at proposals for large-scale human interventions that might help stabilise Antarctic ice sheets, such as the building of undersea berms and walls. The chapter concludes with an analysis of the intersection of the Antarctic geoengineering discourses and Antarctic governance.

Antarctica, climate change and geoengineering discourse

Antarctica is an important site for scientific investigation of global environmental change. The Antarctic cryosphere and Southern Ocean are well recognised as key elements of the Earth system that are central to any investigation of the causes and impacts of global environmental change. Monitoring the concentration of atmospheric ozone levels above the Antarctic region was crucial to the scientific discovery of the problem of ozone-depleting gases. The reduction in concentration of atmospheric ozone and related increase in levels of ultraviolet radiation is primarily found at polar latitudes, including areas of the Antarctic continent and Southern Ocean, during Spring. Similarly, the drilling and extraction of ice cores from the Antarctic ice sheet has been fundamental to investigating the composition of the atmosphere over hundreds of thousands of years (including carbon dioxide concentration) and its relationship to the global climate. The Southern Ocean's East–West circulation pattern and distribution of ocean currents play a significant role in weather and climate patterns across the planet. Human-induced increase in the concentration of carbon dioxide in the atmosphere is also a driver of increased levels of acid-ification found in the waters of the Southern Ocean. The Antarctic cryosphere and Southern Ocean are therefore key sites for investigating and observing the impacts of global environmental change, with climate change being one of the most important of these impacts.

Human governance of such global environmental problems has a mixed record. On the positive side, scientific knowledge of the existence and impact of ozone-depleting gases developed during the late 1970s to early 1980s and was quickly acted upon by the international community in the mid-1980s through the formation of the Vienna Convention for the Protection of the Ozone Layer 1985 and Montréal Protocol on Substances that Deplete the Ozone Layer 1987. These international treaties formed the basis of institutions that have significantly reduced the level of emissions of ozone-depleting substances into the atmosphere and the attendant risks to human health and the natural world. On the other hand, while the causes of human-induced climate change have been recognised in scientific and policy communities since the late 1980s, more modest and lim-ited action in terms of reducing emissions has occurred. The United Nations Framework Convention on Climate Change (UNFCCC) 1992 is now twenty-seven years old, and there have been three major subsidiary treaties or agreements (Kyoto Protocol 1997, Copenhagen Accord 2009 and Paris Agreement 2015). However, global greenhouse gas emissions continue to rise each year and are on track to produce at least three degrees of global warming above pre-industrial levels by 2100. Similarly, international agreements relating to climate change and governance of the oceans (United Nations Convention on the Law of the Sea 1982) are yet to seriously engage with the subsidiary issue of ocean acidification that affects the Southern Ocean.

On one view, the success of the ozone treaties was largely a function of the moderate difficulty of that problem. The ozone-depleting gases which were identified as the key culprits in depletion of the ozone layer were human-cre-ated gases and only developed during the early years of the twentieth century for industrial application in propellants, refrigerated gases and manufacturing. Other less-depleting gases were available as substitutes. Fortuitously, many of the large US corporations which were involved in the production and sale of the ozone-depleting gases also held the rights to the substitute gases, thereby making the path from environmentally damaging to environmentally benign behaviour politically easier for the United States government.

Human-induced climate change, on the other hand, is arguably a much more malign issue for governance. Many of the key states and largest global corporations are heavily dependent upon either the export or domestic use of fossil fuels. Dec-adal-long investments have been made in coal, oil and gas infrastructure which make an economic transition away from those fuels economically and politically difficult. Substitute sources of energy, such as renewables, have until recently been viewed, at least on the short-run cost calculations, as being significantly more expensive, parti-cularly in the absence of an international price on carbon emissions. Major economies and regions that have placed a price on carbon pollution have shown significant difficulties in achieving a price level to provide a strong incentive for key industry players to move away rapidly from investment in fossil fuel investment. Australia is a paradigmatic example of this difficulty, in that it enacted an ambitious carbon pricing scheme at a national level in 2012, only to see it repealed by 2014. This means that a realistic view of international climate change policy must accept that the world will likely move only slowly in reducing emissions and that global temperature increase is likely to be above three degrees later this century.

The relatively poor performance of international governance in reducing greenhouse gas emissions and lessening the trajectory of global temperature increase has enlivened scientific and policy discussion of the possibilities of geoengineering as part of the human response to climate change. The idea of geoengineering to lessen the effects of climate change was initially raised in a 1965 report of President Lyndon Johnson's Science Advisory Committee in regard to adding reflective particles to large areas of the ocean (Nuccitelli 2015); it first appeared in scientific peer-reviewed literature in 1977. However, geoengineering was not seriously discussed in climate change governance literature until the mid-1990s, when US authors, such as game theorist/economist Thomas Schelling (1996) and international lawyer Daniel Bodansky (1996), commented on proposals for stratospheric aerosol injection as a way that humans might mimic the natural analogue of stratospheric volcano eruptions in lowering global temperatures. A key example is the 1991 Mount Pinatubo volcano eruption in the Philippines which, due to depositing large amounts of particulates into the stratosphere, caused the global temperature to be reduced by over 0.5 degrees Celsius for a period of twelve months. This initial policy discussion of solar radiation management came out of North America and was proffered as one possible cost-effective response to climate change for developed countries, which were still in the early years of efforts to reduce greenhouse gas emissions. The discussion of geoengineering in this early period in climate policy centred upon SRM techniques, as a possible alternative to bearing the substantial economic costs of reducing emissions.

Interest in geoengineering spiked significantly in 2006. The Nobel Laureate involved in the discovery of the chemical processes leading to atmospheric ozone depletion, Prof. Paul Crutzen, published an article in the leading journal *Climatic Change* arguing in favour of research into SRM techniques such as stratospheric aerosol injection (Crutzen 2006). The Crutzen article appeared at a crucial time in international climate change negotiations – the lead-up to international negotiations to extend the binding national emissions reduction targets of the Kyoto Protocol. The failure of the 2009 Copenhagen meeting of the UNFCCC to extend the binding national targets of the Kyoto Protocol occurred shortly after the release of a seminal report on geoengineering by the Royal Society (The Royal Society 2009). This Royal Society report built on the 2006 Crutzen article in furthering scientific and policy discussion of geoengineering. The Royal Society report specifically covered both SRM proposals and CDR geoengineering proposals, which are aimed at large-scale removal and long-term storage of carbon dioxide from the atmosphere. In 2015, the US National Academy of Sciences published a two-volume report on 'climate intervention': one volume on SRM proposals, and another on CDR proposals (National Research Council 2015a & 2015b). Since the 2009 Royal Society report, there has been a steady stream of scientific papers on geoengineering proposals, and humanities, law and social science papers on governance of research and possible future use of geoengineering.

International institutions have also started to develop geoengineering governance mechanisms. Meetings of the 1992 UN Convention on Biological Diversity and 1996 London Protocol on Ocean Dumping both developed non-binding resolutions on geoengineering in the late 2000s. In 2013, the London Protocol went further in developing an amendment to govern one type of geoengineering involving the oceans (ocean fertilisation), and a framework with the potential for expansion to govern other types of marine geoengineering techniques in the future. Although not yet in force, this 2013 amendment to the London Protocol represents the first set of binding rules in the international system that is specifically directed towards geoengineering, in that instance, CDR using the oceans.

Most of the scientific work on solar radiation management and carbon dioxide removal has been limited to conceptual development and laboratory modelling. However, there have been some geoengineering field tests and planned experiments over the last twenty years. For SRM, Russian scientist Yuri Izrael claims to have carried out a low-level, small-scale atmospheric test of reflective particles in the mid-2000s (Hamilton 2013, p. 139). Greater public knowledge attaches to a stratospheric aerosol test known as the SPICE experiment, which was planned for the United Kingdom in 2012. The SPICE experiment proposed depositing approximately a bathtub-volume of water droplets in the atmosphere above southern England, at a height of approximately one kilometre, as a testbed for potential later stratospheric aerosol injection activity. However, the SPICE experiment was cancelled during the project, before field testing commenced, due to concerns raised by civil society groups and also intellectual property issues between the researchers involved. For CDR, scientists have conducted field testing in ocean fertilisation since the 1990s. Initially, this work was directed at better understanding the biochemistry of the oceans. However, from the late 2000s, this research has been also directed towards investigating the potential of ocean fertilisation to sequester carbon dioxide for storage in the deep ocean bed. In 2012, American entrepreneur Mr Russ George was involved in an experiment using ocean fertilisation, ostensibly to improve salmon fishery productivity, off the coast of British Columbia. In 2009, a joint German and Indian team of researchers, working on the LOHAFEX experiment, conducted ocean iron fertilisation activities in the Southern Ocean.

Further small-scale geoengineering experiments are also planned and/or in progress. In the next few years, atmospheric physicists at Harvard University plan to carry out a stratospheric aerosol test over an area of New Mexico using about one kilogram of calcium carbonate particles. This project is known as the SCOPEX experiment and will involve a stratospheric weather balloon taking custom-built machinery into the stratosphere to release the small particles of calcium carbonate. The purpose of this experiment will be to analyse the behaviour and reflective properties of the calcium carbonate particles, not to impact the climate per se. The Queensland and Australian governments have recently invested approximately three million dollars in research into techniques that might assist the Great Barrier Reef (GBR) to avoid coral bleaching events during marine heat waves. One technique, field tested in 2019, is a polymer film which floats on the surface of the water and might be used to protect small, high-value areas of the GBR from bleaching. Another technique involves the use of upwelling technology to bring colder water from a depth of ten to thirty metres to cool corals near the surface. The other technique is brightening marine clouds, using seawater aerosols to shade coral reefs from heat (McDonald et al. 2019).

These proposals demonstrate that geoengineering research is moving from conceptual development to small scale field testing. The discourse of geoengineering is also moving quickly in line with these more tangible practices in geoengineering research and field testing. This extension of geoengineering discourse is now significantly implicating the Antarctic cryosphere and Southern Ocean, the topic to which we now turn.

The enhanced reflectivity discourse

Ice-covered areas of the global cryosphere are large and highlight reflective. They are very important to the Earth's overall level of reflective and radiative balance. Loss of ice in the Antarctic, Arctic, Greenland, Himalayas and other ice-covered areas of the planet is significant in that the planet will absorb more heat in the (less reflective) areas of land or water that remain. This is particularly marked in the Arctic, where sea ice has seen a dramatic reduction due to climate change. This has led to development of an 'enhanced reflectivity discourse' concerning ice-covered areas under threat from climate change. This enhanced reflectivity discourse envisions using various SRM techniques to either wholly, or strategically, protect ice-covered areas from further ice loss.

SRM experimentation to protect sea ice has already undergone small-scale field testing. The American organisation Ice911 – which describes its work as 'a Silicon Valley moon shot aiming to stabilize the climate by returning ice in the Arctic' (Ice911, n.d.) – is experimenting with several football-field-sized areas of ice to test the reflective properties of very small, sand-like, floating, hollow glass beads that might be placed strategically on areas of sea ice formation to increase ice reflectivity. The technique seeks to use these small particles to increase the reflectivity of the sea ice surface and thereby reduce the level of sea ice deterioration. Field testing of this technique has been carried out in Alaska, as described on the Ice911 website:

> As the integral part of our three-step process, deploying and testing our material in an Arctic environment is critical for the Ice911 team to gather

data on material effectiveness, rehearse deployments in the Arctic, test its instrumentation, and inform its climate modeling … Our main test site is in Barrow, Alaska located at the Northernmost tip of the U.S. in the Arctic circle, and home to our largest deployments yet. In 2017, we deployed our material solution on 17,500 square meters (over three football fields) of ice, and in 2018, we deployed 15,000 square meters of material with successful results. With our proprietary remote monitoring buoy, we then track how our material is performing during the spring melt from our lab in California.

(Ice911 n.d.)

There has also been earlier small-scale testing of this technique in California and Canada during 2016. The results of the field testing were published in a 2018 article (Field et al. 2018) in the leading journal of the American Geophysical Union, *Earth's Future*, reporting moderately promising initial results.

The Ice911 proposal is not to cover the whole of the Arctic sea with the sand-like reflective glass beads, but rather to focus upon limited areas of sea ice formation where the technique was likely to have the best impact. It is envisaged that this will be achieved by using three to six large container ships to disperse the sand-like glass beads on areas of the ocean at times of the year when sea ice is about to form so as to provide a reflective upper layer on the sea ice (Ice911 n.d.). Importantly, while the Ice911 research carried out to date has centred on restoring Arctic sea ice, the *Earth's Future* paper reporting on the field test results of the Ice911 project specifically raises the possibility of similar techniques being used in the Antarctic to increase the albedo of sea ice areas in the polar South (Field et al. 2018). The enhancing reflectivity discourse evidenced by this type of geoengineering SRM technique thus envisages its use in areas of sea ice formation in the Southern Ocean.

Clouds are also a part of plans for SRM geoengineering. Weather modification programmes aimed at increasing the amount of precipitation over land areas have been in place in the United States, Australia and various other countries since the 1950s. While the effectiveness of these programmes is controversial, the practice of using aircraft to seed clouds with small aerosol particles to increase their precipitative properties is well-established, and soon to be taken to a new scale through large-scale experiments on the Tibetan Plateau in China (Dockrill 2018). Similar proposals have been made for cloud seeding techniques to be used over areas of the Arctic to increase the reflectivity of marine clouds and hence their ability to limit ice loss. Marine cloud seeding would use either ships or aircraft to deposit aerosols into low-lying marine clouds in areas where they would have the best chance of reflecting sunlight away from sea ice and/or land ice areas. Modelling of these techniques is being led by US Department of Energy scientists in the Pacific Northwest (Kravitz 2014). Aside from this modelling work, small-scale field testing of marine cloud brightening occurred in 2011 with the E-PEACE project off the Californian coast. The E-PEACE project involved a small-scale, controlled aerosol perturbation experiment by a research group from the University of Washington in

Seattle. This research group is also proposing to carry out some further field testing and experimentation on marine cloud brightening, using saltwater particles to increase the albedo of marine clouds (University of Washington 2019).

A paper published by this group in the *Philosophical Transactions of the Royal Society of London A* specifically raises the possibility of using marine cloud brightening (MCB) to assist in preventing loss of the West Antarctic Ice Sheet, particularly the Thwaites Glacier.

> The marine ice shelf supporting the Thwaites Glacier is reportedly being eroded from underneath by warmer ocean currents. If MCB proves to be effective, and if the ocean circulation patterns are stable, it may be possible to cool the ocean surface where the warmer currents originate, using MCB in an almost 'surgical' manner at local or regional scales... . The arguments just presented must be considered provisional and further research is required in order to determine whether MCB might be capable of producing a regionally directed cooling, either alone or in concert with other global measures capable of restoring stability to Antarctic ice sheets.
>
> (Latham et al. 2014)

If marine cloud brightening techniques, after further modelling and field testing, prove to have some success in increasing albedo of ice-covered areas, they could have potential application in and around the Antarctic continent.

Finally, as discussed above, there is longstanding evidence within geoengineering discourse of the idea of using global-scale stratospheric aerosol injection to replicate the atmospheric albedo modification properties of stratospheric volcanoes. As discussed above, the Crutzen (2006) article re-enlivened interest in stratospheric sulphate aerosol injection techniques in an effort to lessen the prospects of large-scale climate change impacts, such as destabilisation of the West Antarctic and Greenland ice sheets, which would have a significant effect on global sea-level rise. The enhancing reflectivity discourse, extending to the global scale in this way, has given rise to hopes that global-scale albedo modification through stratospheric aerosol injection may be able to buy time to save these globally significant ice sheets, whilst humanity scales back greenhouse gas emissions. However, enthusiasm for reliance upon global-scale reflectivity enhancements to slow or save areas such as the West Antarctic Ice Sheet has recently suffered. Work by McCusker, Battisti and Bitz (2015) and Applegate and Keller (2015) suggests that, even if stratospheric aerosol injection has some effect in other areas of the planet, it is not likely that it will counteract continued upwelling of warm water in proximity of ice shelves, especially those in the vicinity of the already unstable Pine Island Glacier of West Antarctica. The global enhanced reflectivity discourse on preventing ice sheet loss and significant sea-level rise has therefore recently been tempered, but still remains.

The enhanced carbon sequestration discourse

The planet's oceans moderate global climate change through their ability to absorb heat and carbon dioxide from the atmosphere, thereby reducing the atmospheric effects of increased carbon emissions from human activities. Globally, since the 1960s, the oceans have taken up approximately ninety per cent of the extra heat generated by climate change in the Earth system and nearly thirty per cent of total carbon dioxide emitted from fossil fuel use, cement production and land use change (ACE CRC 2011). Between 2000–2008, the planet's oceans absorbed approximately one quarter of all human carbon dioxide emissions (Global Carbon Project 2010). The Southern Ocean has a crucial role in the global storage of heat and carbon, as the area below 30° South absorbs approximately forty per cent of the total global ocean uptake of anthropogenic carbon dioxide (ACE CRC 2011). This is due to the unique ocean currents of the Southern Ocean, which flow in an anticlockwise motion from West to East around Antarctica and overturn water from deep to shallow parts of the Southern Ocean, and vice versa. The balance between this upwelling and downwelling of the waters in the Southern Ocean regulates the very significant carbon sequestration properties of the region (ACE CRC 2011).

Attention has turned within carbon dioxide removal geoengineering research to whether human interventions might be used to enhance the Southern Ocean's carbon sequestration properties. One such proposal is ocean fertilisation, which aims to remove carbon dioxide from the atmosphere by artificially adding nutrients (such as iron, nitrogen or phosphorus compounds) into the ocean to stimulate growth of marine phytoplankton. When these marine phytoplankton die, some will sink to the ocean floor as sequestered carbon, where it might remain out of contact with the atmosphere for hundreds, or even thousands, of years (ACE CRC 2016). Ocean fertilisation occurs quite naturally in areas of ocean adjacent to large landmasses, where winds can blow dust from the land and spur phytoplankton blooms. The Southern Ocean, being removed from the proximity of significant landmasses, is relatively low in iron content, and so a prime location for ocean fertilisation.

As stated above, ocean fertilisation has been carried out in fourteen scientific field experiments commencing in 1994, with many carried out in the Southern Ocean. In all but one of these experiments, the fertilising agent which was added to the ocean was iron based. As the ACE CRC reported in 2016:

> All but one of the 14 experiments to date have added iron (the exception added phosphorous), and all but one of the iron additions have observed increased growth rates of phytoplankton. Almost all have also shown increases in phytoplankton stocks and carbon fixation, promoting carbon dioxide drawdown into the ocean from the atmosphere by gas exchange. Some of the artificially induced blooms of phytoplankton extended to nearly 1000 km^2 in area and were visible to satellite-based ocean colour sensors … Thus, a major achievement has been the conclusive

demonstration that ocean scarcity of iron controls biological production in nearly one third of the global ocean.

(ACE CRC 2016, p. 11)

Although increased phytoplankton growth was observed in most experiments, the actual increase in carbon exported for storage on the ocean floor was more difficult to ascertain. This was in part because the 'duration of experiments was generally too short to address the associated food web processes ... but also because export processes are complex, time-varying, and often de-coupled from production' (ACE CRC 2016, p. 11). Estimates vary widely (i.e. by two orders of magnitude) on the additional carbon that is exported from surface waters into the deep ocean for a given addition of iron (p. 11). These uncertainties in the success of iron fertilisation can be lessened only through larger scale outdoor experimentation, which would involve greater risks of ecosystem impacts, and are hence yet to be advanced.

The continental margin areas that lie off the landmass and edge of the ice sheets of Antarctica are generally wide, deep and rich in marine life. Most of the marine species living in these areas are in the benthos or seabed and play a vital role in global carbon sequestration. The Antarctic Seabed Carbon Capture Change Project (ASCCC) explains:

> Benthos commonly comprise echinoderms (sea stars, brittlestars, sea urchins), molluscs (clams & snails), corals, sponges, crustaceans, bryozoans (sea mosses) and many other animal types. They eat plankton (such as microscopic plants and animals). Carbon is transported through the system by being fixed in photosynthesis by the tiny algae, which are eaten by benthos, and then buried when the benthic animals die.
>
> (ASCCC 2019)

The ASCCC explains how sequestration of carbon dioxide in Antarctic benthos is increasing as sea ice retracts, thereby acting as a negative feedback on global climate change:

> In 2015 we estimated the carbon stocks and annual increments around the West Antarctic seas. We found that this value has nearly doubled in the last 25 years, in response to sea ice losses. So although rising CO_2 in the atmosphere has driven global warming, which has reduced Arctic and west Antarctic sea ice through warming air and/or sea temperatures, it has led to more carbon accumulation in animals on the seabed (thus less in the air) as a feedback working against climate change.
>
> (ASCCC 2019)

This raises the possibility of artificial ocean fertilisation being used in the future in areas where sea ice is retracting to stimulate phytoplankton blooms and further increase benthic sequestration of carbon. While this does not appear to be have been taken up in the literature to date, the significant negative feedback provided by benthic carbon deserves consideration, not only for protection (from fishing and other human activities), but also for possible enhancement through ocean fertilisation. The Antarctic region is therefore not only an important natural sink for carbon dioxide, but also a possible place for enhanced carbon sequestration, which might, given appropriate further research, see carbon stored at large scale on the seafloor and in benthic marine life.

The glacial stabilisation discourse

The third Antarctic geoengineering discourse does not fit neatly within either the SRM or CDR categories, as it involves neither reflectivity modification nor sequestering carbon from the atmosphere. While this group of proposals has been referred to as forms of 'geoengineering', they are in effect mega-scale human engineering proposals to modify the local environment of the cryosphere in an attempt to buttress or stabilise glaciers. For example, Moore et al. (2018, p. 304), in the journal *Nature*, introduce these ideas as follows:

> Is allowing a 'pristine' glacier to waste away worth forcing one million people from their homes? Ten million? One hundred million? Should we spend vast sums to wall off all the world's coasts, or can we address the problem at its source? Geoengineering is a political and societal choice, because people's reactions depend on how the issue is framed. Buttressing of glaciers needs a serious look. It should have fewer global environmental impacts than other proposals being discussed for reducing sea-level rise, such as injecting aerosols into the stratosphere to reflect sunlight and cool the planet.

One idea put forward in this article is 'blocking warm water' by the construction of undersea walls and berms in areas just off glaciers (the Jakobshavn Glacier in Western Greenland is offered as a potential example) to reduce the passage of destabilising warm water under them. Another idea, suggested for the Pine Island and Thwaites Glaciers in West Antarctica involves 'supporting ice shelves' by artificially pinning the ice shelves in front of the glaciers by constructing berms and artificial islands, which would extend from outcrops or be built on the sea floor (Moore et al. 2018, p. 304). It is acknowledged that this would be a practically more ambitious and difficult task:

> Material could be shipped to Antarctica from elsewhere in the world, or dredged or quarried locally. But it would be difficult in practice for engineers to work around the ice shelves, which grow and shrink as the glaciers, sheets and conditions fluctuate. Sea ice would also get in the way. Technologies might need to be developed to operate beneath floating ice. Major disturbances to local ecosystems would be expected and would require thorough assessment before and after pinning.
>
> (Moore et al. 2018, p. 304)

Moore et al. (2018) have also explored the possibility of drying sub-glacial streams by drilling into areas of ice sheets and pumping such water to the surface to refreeze it. These sub-glacial streams facilitate ice flow, so their removal could assist with stabilisation of Antarctic ice sheets. These ideas are likely to be extremely costly to implement; however, Moore et al. point out that the very significant financial costs of such work would be small compared with the global financial costs from sea-level rise, if there is significant loss of the Antarctic or Greenland ice sheets. The proposals for this work are not to attempt to stabilise all the ice sheets, but only in strategic areas that might be expected to yield the best results.

The construction of artificial islands or berms and walls several kilometres in length in undersea areas off the Antarctic ice sheets would involve very significant transfer of materials and construction infrastructure from outside the Antarctic region. It would also necessarily involve significant machinery and personnel for construction and the attendant environmental risks of large numbers of people living in these areas during the construction period. There would also be significant environmental risks of oil spills and other pollution during the construction phase. The effectiveness of these strategies would largely be uncertain until they were constructed. For these reasons, the Antarctic glacial stabilisation discourse might be viewed by some as dystopian thinking. However, the onset of climate change impacts, particularly sea-level rise, over coming decades, might see these types of ideas being looked at more realistically.

What the Antarctic glacial stabilisation discourse does show, however, is that current thinking is moving well beyond the usual consideration of the environmental impacts of human activities in Antarctica. There is a realisation of the profound global forces of human intervention that will shape the Antarctic cryosphere and the possibilities of human action to protect both the Antarctic environment from radical change and preserve the coastlines of the world's major cities.

Geoengineering discourse and Antarctic governance

It is clear from the above analysis that the frontiers of human thought and expectations, as expressed in scientific literature, now contain several ways in which geoengineering discourse implicates the Antarctic cryosphere and Southern Ocean. Through the enhanced reflectivity discourse, the Antarctic cryosphere is significantly implicated in possible use of SRM techniques, such as reflective, sand-like particles or marine cloud brightening, that might increase reflectivity of areas in and above the ice, and thereby lessen their deterioration. Similarly, on a global scale, stratospheric aerosol injection proposals have raised the possibility of lessening rates of ice-sheet loss. The Southern Ocean and Antarctic region play an important role in the global carbon cycle. In the enhanced carbon sequestration discourse, the Southern Ocean is significantly implicated in proposals for large-scale CDR through ocean fertilisation. Further, the retraction of sea ice has given rise to increased carbon sequestration in benthic areas of the coastal margin of Antarctica. This is another major sequestration function of the Antarctic region and Southern Ocean which not only requires protection by the Antarctic Treaty System (ATS) from potentially damaging human activities in the region, but also raises the possibility of enhancement through ocean fertilisation activities. The glacial stabilisation discourse proposes to most directly affect the Antarctic region through the construction of mega-engineering projects to help stabilise glaciers and ice sheets and lessen sea-level rise.

The contributions to this volume by Chaturvedi, Stephens, Hemmings and Salazar destabilise the notion of separateness between human activity and Antarctica. It may well be that Antarctica, as an Eden of natural, untouched creation was lost some time ago. However, these three geoengineering discourses demonstrate that any notion of separateness between human activity and the Antarctic region is being further destabilised. The geographical remoteness and climate extremes of Antarctica have in the past created a barrier to human interaction with the continent. However, it is clear that climate change as a driver of global environmental change, including significant warming of the planet, is breaking down further any barriers between the Antarctic region and the rest of the planet. The Antarctic geoengineering discourses are further breaking down those barriers by constructing the possibility that the human footprint on Antarctica will, in the coming decades, extend beyond the usual scientific research, tourism and marine resource extraction to activities designed specifically to lessen the impacts of global climate change.

It is also clear from the governance perspective that any exceptionalism of the Antarctic Treaty System as the primary institution for governing the Antarctic cryosphere and Southern Ocean is likely to be lessened by any of the geoengineering proposals discussed here being put into place. The 1996 London Protocol on Ocean Dumping was developed by states as a treaty on ocean dumping that applies across all oceans, including the Southern Ocean. The London Protocol has developed specific rules around marine geoengineering, particularly ocean fertilisation, that have been developed outside the Antarctic Treaty Consultative Meeting process. Similarly, the 1992 United Nations Convention on Biological Diversity has sought to prohibit geoengineering activities, except for small-scale scientific research. The governance forums that have engaged with geoengineering activities to date are therefore primarily those located outside the Antarctic Treaty System. If global approaches are developed on geoengineering governance, they are more likely to emerge out of the Intergovernmental Panel on Climate Change, United Nations Framework Convention on Climate Change, London Protocol or United Nations Environmental Programme, than the Antarctic Treaty System. For that reason, the above Antarctic geoengineering discourses raise the importance of future interaction between the ATS and other resource and natural resource management regimes operating in the region (McGee & Haward 2019).

That is not to say that the Antarctic Treaty System, particularly the 1991 Madrid Protocol on Environmental Protection and 1980 Convention on Antarctic Marine Living Resources, could not take an important role in managing geoengineering research and field testing, at least for states that are parties to these treaties. However, for countries that are outside the Antarctic Treaty

System there will more likely be recourse to these wider environmental and natural resource management regimes related to ocean dumping, biodiversity protection, climate change and other issues. In this sense, the geoengineering discourses for Antarctica may well become another pressure point whereby the exceptionalism of the Antarctic Treaty System, in seeking to comprehensively manage the region, will be put under a further pressure. In this way, the Antarctic geoengineering discourses illustrate that the continent is not only a 'Lost Eden' due to increased biophysical presence of humans, but also because the ATS may no longer comprehensively govern issues affecting Antarctica.

Conclusion

Climate change is a driver of large-scale, global biophysical change that is impacting the Antarctica cryosphere and Southern Ocean and will change these regions in profound ways. It is therefore not surprising that the discourses of geoengineering have engaged with the Antarctic cryosphere and Southern Ocean, because these regions are vital for the operation of the global climate and have been significantly implicated in proposals for enhanced reflectivity on ice area, enhanced carbon sequestration and glacial stabilisation. However, the product of the extension of these geoengineering discourses to Antarctica is likely to be a continued reduction in the barriers of both conceptual and practical separateness of Antarctica from other parts of the globe and other elements of the global governance system. This chapter therefore highlights that, through climate change, the Anthropocene is reshaping Antarctica in both a material and a discursive sense.

Note

1 'Antarctic continent' is used here to include areas of land, ice sheets and ice tongues, but not the seasonal sea ice formations.

References

Antarctic Climate & Ecosystems Cooperative Research Centre (ACE CRC) 2011, *Position Paper: Climate Change and the Southern Ocean*, The Antarctic Climate & Ecosystems Cooperative Research Centre, Hobart, viewed 20 March 2019, <http://a cecrc.org.au/wp-content/uploads/2015/03/2011-ACE-PA-Climate-Change-a nd-the-Southern-Ocean.pdf>.

ACEC CRC 2016, *Position Analysis; Ocean Fertilisation*, The Antarctic Climate & Ecosystems Cooperative Research Centre, Hobart, viewed 20 March 2019, <http://a cecrc.org.au/wp-content/uploads/2016/07/ACE106_Position-Analysis_Ocean-Fert_ April-2016_WEB.pdf>.

ACE CRC 2017, *Position Analysis: The Antarctic Ice Sheet and Sea Level*, The Antarctic Climate & Ecosystems Cooperative Research Centre, Hobart, viewed 20 March 2019, <http://acecrc.org.au/wp-content/uploads/2017/10/2017-ACECRC-Positio n-Analysis_Ice-Ocean-Interaction.pdf>.

ASCCC 2019, viewed 20 March 2019, < https://www.asccc.co.uk/our-project/>.

Applegate, PJ & Keller, K 2015, 'How effective is albedo modification (solar radiation management geoengineering) in preventing sea-level rise from the Greenland Ice Sheet?', *Environmental Research Letters*, vol. 10, pp. 1–10.

Bodansky, D 1996, 'May we engineer the climate?', *Climatic Change*, vol. 33, pp. 309–321.

Brooks, ST, Jabour, J, van den Hoff, J & Bergstrom, DM 2019, 'Our footprint on Antarctica competes with nature for rare ice-free land', *Nature Sustainability*, vol. 2, pp. 185–190.

Crutzen, P 2006, 'Albedo enhancement by stratospheric sulfur injections: a contribution to resolve a policy dilemma?', *Climatic Change*, vol. 77, pp. 211–219.

Dockrill, P 2018, 'China's "sky river" will be the biggest artificial rain experiment ever', *Science Alert*, 28 April 2018, viewed 20 March 2019, <https://www.sciencealert.com/how-china-s-sky-river-will-be-the-biggest-artificial-rain-experiment-ever-cloud-seeding>.

Dryzek, JS 2006, *Deliberative global politics: discourse and democracy in a divided world*, Polity Press, Cambridge.

Field, L, Ivanova, D, Bhattacharyya, S, Mlaker, V, Shotlz, A, Decca, R, Manzara, A, Johnson, D, Christodoulou, E, Walter, P & Katuri, K 2018, 'Increasing Arctic sea ice albedo using localized reversible geoengineering', *Earth's Future*, vol. 6, no. 6, pp. 882–901.

Global Carbon Project 2010, *10 Years of Advancing Knowledge on the Global Carbon Cycle and its Management*, viewed 14 March 2019, <https://www.globalcarbonproject.org/global/pdf/GCP_10years_med_res.pdf>.

Hamilton, C 2013, *Earthmasters: the dawn of the age of climate engineering*, Yale University Press, London.

Ice911 n.d., viewed 6 April 2019, <http://www.ice911.org/>.

Kravitz, B 2014, 'The bright side of Arctic clouds', Research Highlights: US Department of Energy, Office of Science, 3 December 2014, viewed 20 March 2019, <https://climatemodeling.science.energy.gov/research-highlights/bright-side-arctic-clouds<https://

Latham, J, Gadian, A, Fournier, J, Parkes, B, Wadhams, P & Chen, J 2014, 'Marine cloud brightening: regional applications', *Philosophical Transactions. Series A, Mathematical, Physical, and Engineering Sciences*, vol. 372, no. 2031, 20140053, doi:10.1098/rsta.2014.0053.

McCusker, KE, Battisti, DS & Bitz, CM 2015, 'Inability of stratospheric sulfate aerosol injections to preserve the West Antarctic Ice Sheet', *Geophysical Research Letters*, vol. 42, pp. 4989–4997.

McDonald, J, McGee, J, Brent, K & Burns, W 2019, 'Governing geoengineering research for the Great Barrier Reef', *Climate Policy*, doi:10.1080/14693062.2019.1592742.

McGee, J & Haward, M 2019, 'Antarctic governance in a climate changed world', *Australian Journal of Maritime and Ocean Affairs*, vol. 11, no. 2 [forthcoming].

Moore, J, Gladstone, R, Zwinger, T & M Wolovick 2018, 'Geoengineering glaciers to slow sea level rise', *Nature*, vol. 555, pp. 303–305.

Nash, KL, Cvitanovic, C, Fulton, EA, Halpern, BS, Milner-Gulland, EJ, Watson, RA & Blanchard, JL 2017, 'Planetary boundaries for a blue planet', *Nature Ecology & Evolution*, vol. 1, pp. 1625–1634.

National Research Council 2015a, *Climate Intervention: Reflecting Sunlight to Cool Earth*, The National Academies Press, Washington, DC, doi:10.17226/18988.

National Research Council 2015b, *Climate Intervention: Carbon Dioxide Removal and Reliable Sequestration*, The National Academies Press, Washington, DC, doi:10.17226/18805.

Nuccitelli, D 2015, 'Scientists warned the US president about global warming 50 years ago today', *Guardian*, 5 November 2015, viewed 6 April 2019, <https://www.theguardian.com/environment/climate-consensus-97-per-cent/2015/nov/05/scientists-warned-the-president-about-global-warming-50-years-ago-today>.

Rockström, J, Steffen, W, Noone, K, Persson, Å, Chapin, FS, Lambin, E, Lenton, TM, Scheffer, M, Folke, C, Schellnhuber, H, Nykvist, B, De Wit, CA, Hughes, T, van der Leeuw, S, Rodhe, H, Sörlin, S, Snyder, PK, Costanza, R, Svedin, U, Falkenmark, M, Karlberg, L, Corell, RW, Fabry, VJ, Hansen, J, Walker, B, Liverman, D, Richardson, K, Crutzen, P & Foley, J 2009, 'Planetary boundaries: exploring the safe operating space for humanity', *Ecology and Society*, vol. 14, no. 2, article 32.

Schelling, T, 1996, 'The economic diplomacy of geoengineering', *Climatic Change*, vol. 33, pp. 303–307.

The Royal Society 2009, *Geoengineering the climate: science, governance and uncertainty*, RS policy document 10/09, The Royal Society, London.

Thiele, LP 2019, 'Geoengineering and sustainability', *Environmental Politics*, vol. 28, no. 3, pp. 460–479.

University of Washington 2019, *Marine Cloud Brightening Project*, viewed 20 March 2019, <http://mcbproject.org/index.html>.

Young, O 2017, *Governing complex systems: social capital for the Anthropocene*, MIT Press, Cambridge, Mass.

5 The Anthropocene melt

Antarctica's geologic politics

Juan Francisco Salazar

Introduction

The Anthropocene has become the impulse through which a large number of disciplines across the academy are appraising, debating or redefining conceptions of nature–culture. Antarctica offers a distinctive way of approaching this concept: the region has always presented itself as an inherently futures-oriented problem and a serious test for humanity's coordinated capacity to exercise foresight (Salazar 2015). This involves not only protecting the region's fragile ecosystems, but also rethinking our species as part of (and in relation with) nature, and mobilising novel experiments with living differently in the Anthropocene. In this short essay, I discuss in broad terms how Antarctica is an important object through which to think the Anthropocene.

In May 2016, an international interdisciplinary symposium titled *Antarctica in/and the Anthropocene* was held at the Department of Anthropology, Pontificia Universidad Católica in Santiago, to coincide with the Thirty-ninth Antarctic Treaty Consultative Meeting held in Chile. This event brought together, arguably for the first time, a range of perspectives from the humanities and social sciences with a range of Antarctic geosciences to exchange ideas about Antarctica as a crucial object for thinking the Anthropocene. Drawing on an interdisciplinary impetus, the participants at this event were invited to open a dialogue around the lessons that might be learnt from Antarctica in thinking about life in the Anthropocene, an epoch where human and Earth futures are increasingly entangled and interdependent in their mutual uncertainty. In part, the underlying motivation for this event was to respond to how scientific formalisation of the Anthropocene seems to be characterised by a tension between solid surfaces and fluid media (Simonetti & Ingold 2018) that sets geophysical sciences and humanities scholars apart.

Over little more than two centuries, human activities have transferred to the atmosphere, in the form of gases or heat, a substantial part of the hydrocarbons that took millions of years to accumulate on Earth. More substantially and worryingly, the Anthropocene has come to signal a time interval in which nowhere – neither the furthest reaches of the stratosphere nor the lowest point in the marine abyss – is undamaged by the doings and silt of humankind. Bronislaw Szerszynski (2012, p. 169) puts it succinctly: '[the] truth of the

Anthropocene is less about what humanity is doing, than the traces that humanity will leave behind'. 'Humanity' here needs to be read with caution; as Madelaine Fagan (2019, p. 55) notes, these assumptions about the Anthropocene as all of humanity's doings 'rely on a transposition of geological onto historical and political periodization'.

As anthropologists Cymene Howe and Anand Pandian observe, the idea of the Anthropocene has spread astonishingly quickly, 'dislodging familiar terms like nature and environment from their customary preeminence as signs of the world beyond ourselves' (2016, n.p.). While the geophysical sciences in Antarctica have, for some time now, provided unequivocal evidence of the human impact of a geological transition at a planetary scale, humanities scholars working in and on Antarctica have only recently begun to address the question of an Anthropocene Antarctica (Zarankin & Salerno 2014; Salazar 2015).

As has been hotly debated, the Anthropocene has come to express the current time interval as a new geological epoch defined by human agency where human activity has so profoundly impacted geology and atmospheric cycles that a new geological unit has become necessary to account for measuring its impact on Earth systems. In some cases, this impact has also extended to invasive species in new habitats, intensive agriculture and soil depletion, urbanisation, pollution, plastic waste and global warming, among other anthropogenic forces and processes that began accelerating after the 1950s. Several Earth systems scientists have been arguing for some time that the driving forces pushing this 'Great Acceleration' constitute an interlinked system, characterised by population growth, increased consumption, the abundance of cheap energy and the implementation of liberalising economic policies (Steffen et al. 2015). While the term Anthropocene risks neglecting the presence – and force – of terrestrial processes that exist independently of human relationships, we must acknowledge that what is at stake here and now is the predicament that humanity is no longer able to control most of the feedback effects derived from its own actions. As Nigel Clark (2014, p. 25) rightly observes, 'what is vital for critical thinkers in the humanities and social sciences to recognize … is that the scientific thematization of the Anthropocene is as much about the decentring of humankind as it is about our rising geological significance'. As discussed earlier, the Anthropocene logic tends to represent humans as a single entity – that is, the human species as a geophysical force (Bauer & Ellis 2018, p. 210) – which in turn raises fundamental epistemological and ontological questions about the 'who' and 'when' of the Anthropocene (Fagan 2019). Kathryn Yusoff (2018, p. xiii) extends this critique to talk about the 'racial blindness of the Anthropocene … that permeates its comfortable suppositions and its imaginaries of the planetary'.

Characterisations of this new period proliferate: the Anthropocene as aftermath (of capitalism and late liberalism); the Anthropocene as the disaster to end all disasters; the Anthropocene as transition. As these multiple interpretations suggest, it is difficult to theorise an epoch of this magnitude while it is still unfolding. But perhaps this time of transition can also be described in an original way: in his novel *2312* (2012), US science fiction writer Kim Stanley Robinson's fictional character,

Charlotte (a historian) develops a periodisation from the early twenty-first century to the early twenty-fourth century. She calls the period from 2005 to 2060 'The Dithering'. The Cambridge Dictionary defines 'to dither' as being unable to make a decision about doing something; Robinson uses it to describe our current epoch as 'a state of indecisive agitation' over climate change and pre-empting a period called *The Crises* (2060–2130) (Robinson 2012, p. 144–145; see also Haraway 2015). Robinson's term is one of the best ways I have found to qualify these times of perplexity, where humanity (or rather, its political leaders and economic elites) finds itself dithering.

This is a compelling reason to think – and act on – the idea that the Antarctic provides a unique opportunity to develop an affirmative relationship across disciplinary divides. The continent offers itself not only as a laboratory for science, but also a laboratory for thinking about alternative ways of living in the Anthropocene. It also creates a space for dialogue about what sort of ethics this new thinking might require. The scientific study of the impacts of global ecosystems change on Antarctica, and the role that Antarctica in turn plays in shaping global ecosystems change, is not new. As a plethora of research shows, the Antarctic continent and surrounding ocean are undergoing a profound transformation, impelled largely by accelerated change in its ecosystems dynamics. Scientists paint a sobering picture of an unfolding and relentlessly unravelling future where changes will only intensify over the next fifty years. These changes are also linked to shifting geopolitical undercurrents, improved technological and logistical capabilities, intense human activities on the continent and in the surrounding ocean, and increased interest in its bio-resources. The scope of human activities in the southern polar region has changed dramatically over the last hundred years: Antarctica is becoming an 'anthropogenic landscape' (Glasberg 2012, p. xxvii), where the challenges of intensifying human activities mean that the current governance system may be insufficient to meet the environmental protection obligations set out under the Madrid Protocol twenty-five years ago.

Thinking about the Anthropocene suggests a complex blend of sociopolitical and physico-material negotiations. The Anthropocene is a category that geologises human existence as it 'punctuates the most massive time scale in our lexicon: the geo-logic' (Howe 2016, n.p.). To continue thinking with Howe:

> the great melting at the top of the world, and the bottom as well, has us wondering about the cool, ancient time that is being washed away. It is as though we are living in the Reformation. Earth's cryosphere is sloughing off as we watch in real time, turned to mush and puddles. One begins to sense a growing nostalgia for the deep history ice holds. It has us thinking *cene*-icly – backward and forward – as we face a precarious future compounded by generational effects and wicked ethics.
>
> (Howe 2016, n.p.)

From a different perspective, Clark (2014, p. 19) observes how 'the current problematization of planetary "boundary conditions" is indicative of the need for new ethical engagements, but is also suggestive of a new kind of "geologic politics" that is as concerned with the temporal dynamics and changes of state in Earth systems as it is with political issues revolving around territories and nation state boundaries'. In this sense, in thinking Antarctica *and* the Anthropocene, the questions that arise relate to the kind of lessons that can be learned from Antarctica that are useful and important to take into account when grasping planetary changes. In the next section, I provide a short critical engagement with how a geo-politics of the Anthropocene plays out in an Antarctic context. I then move on to focus on the melting of the cryosphere as a material-semiotic device to discuss the importance of Antarctica in the Anthropocene.

Geologic politics in Antarctica

> Our drowned cities ... would begin to be covered by sand, silt, and mud, and take the first steps towards becoming geology. The process of fossilization will begin.
>
> (Zalasiewicz 2008, pp. 84–85)

As historian Dipesh Chakrabarty (2009) argues, the Anthropocene marks the end of the binary distinction between humanist and scientific knowledge that underpinned the Enlightenment rationalist project and marks a need to establish a radically different kind of politics. Hemmings et al. (2017, p. 3) make it clear that, for a continent that has historically been deemed as 'essentially separate from the global political and economic system', Antarctica has always posed and continues to pose a complex socio-political question. They put it this way: 'the continent for science and peace is also a continent created by politics, maintained by politics – and indeed, generating politics' (p. 1). So, if Antarctica is generative of politics, and the Anthropocene requires a new kind of politics, it is unexpected that, in their introduction to their *Handbook to the Politics of Antarctica*, Hemmings et al. do not allude to Antarctica in the context of an 'Anthropocene politics'. If the Anthropocene marks a reawakening, triggered not so much by nostalgia for a more stable past, but by a sense of political urgency about an uncertain future, as Simonetti and Ingold (2018) argue, then a geopolitics of Antarctica is inseparable from a geological politics of Antarctica. In 'The Antarctic Treaty System and the Anthropocene', Tim Stephens argues that, despite Antarctica's historical and biogeographical isolation, 'the Anthropocene's signature is inscribed deeply there, from the ozone hole etched in the southern sky to the cleaving of the ice shelves into the Southern Ocean' (2018, p. 29). There cannot be any serious discussion about Antarctic politics in the Anthropocene without engaging with what Stephens calls 'the global forces let loose by human hands' (p. 29) that are increasingly impacting Antarctic ecosystems. This has implications for the critical legal imaginings of Antarctica and the Southern Ocean, as well as for a novel 'geo-logic' reimaging of the southern polar region.

The Arctic and parts of the Antarctic Peninsula are warming at twice the rate of the rest of the planet. In both instances, they have become sources of imagery of amplified environmental change. They have also turned into a spatial setting for climate crisis discourses (Paglia 2016) and – in the case of the Arctic – an opportunity to expand economic exploitation. As briefly discussed in this chapter, ice indeed plays an increasing role in identifying and defining the Anthropocene. The recurrence of northern hemisphere glaciation and the stability of the Greenland Ice Sheet are both potentially vulnerable to human impact on the environment. The Greenland and Antarctic Ice Sheets have not yet created large changes in the landscape or sea level, but some projections suggest that they might in the next few centuries. In this chapter, I discuss the relevance of an emergent geologic-politics of the Anthropocene in Antarctica, understood as a complex blend of socio–political and semiotic–material negotiations, where heightened geopolitical interest in these regions and their resources is continually countered by increasing calls for the protection of polar ecosystems, and for reflecting upon the futures associated with the advent and expansion of the Anthropocene in these regions.

The melting of the cryosphere: ice as the 'stuff of time'

The cryosphere is waning. Glaciers are retreating. Icebergs and ice sheets are calving. Ice shelves are collapsing. Permafrost is thawing. Ice contains stories of Earth from several million years ago. Following Tim Ingold's argument (2012, p. 41) that materials are not only in time, but are the 'stuff of time' itself, we might say, then, that ice is in a permanent process of becoming time (Salazar 2018; Simonetti & Ingold 2018). Ice as a material opens new questions and opportunities, not only for the scientists who study its dynamics and its trapped bubbles of air, but also for the humanities scholars who think about how the Anthropocene comes to signal a time interval in which the furthest reaches of the Earth, from the stratosphere to the deep sea, are affected by the actions and detritus of humankind. In his epitaph to the Holocene, journalist Mike Davis (2008, n.p.) wrote: 'our world, our old world that we have inhabited for the last 12,000 years, has ended, even if no newspaper in North America or Europe has yet printed its scientific obituary'. More than a decade on, that obituary is yet to be written: the Working Group on the Anthropocene of the International Commission on Stratigraphy is seemingly reluctant to formalise the Anthropocene as a geological unit within the Geological Time Scale.

There is a lack of consensus on the best place on Earth to mark the Anthropocene's dawn. The topic continues to be a hotly contested issue that brings about occasional controversy among geologists and palaeo-ecologists, who continue to comb the planet for an unambiguous location to place a 'golden spike' that would define the beginning of the Anthropocene. After two decades of debate about the geological strata, it is possible that the dawn of the Anthropocene could instead be marked by a chemical or biological signal. Whether it be the surge in atmospheric radioactivity from atomic bomb tests in the 1950s (Waters et al. 2015), an upsurge in the concentration of microplastics in the environment or changes in the human gut microbiome, a golden spike requires a signal to be recorded in some kind of material that was accumulating when the Anthropocene began.

In 2008, a golden spike for the Holocene was marked from an ice core extracted from the Greenland Ice Sheet and is now stored at a facility in Copenhagen. This was the first time that a golden spike had ever been marked from ice. So, would ice stratigraphy from Antarctica prove useful to mark changes in atmospheric chemistry linked to human activity? The answer is most likely yes. But, considering that ice sheets are a major facet of the Earth system and one of the most astonishing signs of global climate visible from outer space, it is surprising that the existence, evolution and melting of ice sheets have not played a major role in defining the Anthropocene as a new geological epoch. Nonetheless, signals and geological signatures evident in ice cores are becoming increasingly important. Human influence can clearly be discerned in several ice-core measurements (Wolff 2014), from sharp spikes in atmospheric radioactivity from nuclear testing, to unprecedented increases during the Holocene in concentrations of sulphate, nitrate and metals in the Greenland Ice Sheet, to the appearance in ice core air bubbles of previously undetectable compounds or the exceptional rise in concentrations of carbon dioxide and methane (Wolff 2014).

Antarctic scientists are painting a dire picture of an unfolding and relentlessly unravelling future, where changes will only intensify over the next fifty years. In mid-2018, there was worldwide media coverage of new research which claims that the rate of Antarctic ice melt has tripled in the past five years, with more than two hundred billion tonnes of ice flooding into the oceans annually. In August 2017, glaciologists drilled an ice core in Allan Hills, East Antarctica, which yielded a 2.7-million-year-old ice sample, the most ancient evidence of Earth's atmosphere to date (Voosen 2017). Only a few weeks earlier, a trillion-tonne, 5,800 km^2 iceberg had calved off from the Larsen C Ice Shelf. While this calving event was not the largest on record, it confirmed that Larsen C is now at its smallest extent since the end of the last Ice Age, some 11,700 years ago (Hogg & Hilmar Gudmundsson 2017). More than ten other shelves, further north along the Peninsula, have either collapsed or retreated significantly in recent decades. If we consider for a moment that the Earth's icy polar regions are unique in the Solar System, the drama of this cannot be understated. Furthermore, the accelerated melting of the ice sheets in Greenland and the Antarctic in recent decades has been linked to changes in the rotational axis of the planet, which scientists are calling 'climate-driven polar motion' (Adhikari & Ivins 2016, p. 1). Whether this has happened before in Earth's deep history or it is happening now for the first time is not the point. What is at stake here is that human activities have acquired a force capable of sparking these massive changes to Earth systems, profoundly impacting geological and atmospheric cycles.

It is difficult to avoid the extinction narrative when addressing the unfolding drama of the partial disappearance of the cryosphere. Clark (2010) has a point when he argues that existing social theoretical engagements with physical agency are insufficiently geared towards dissonant or disastrous physical events. The extinction (retreat, thawing, melting and ultimately disappearance) of glaciers inescapably comes to signify a unique 'cryo-historical moment', as historian Sverker Sörlin (2015, p. 327) has called the current human-induced retreat of the cryosphere. This cryo-historical moment of massive deglaciation of the planet can also be seen as a sign of what political theorist William Connolly (2013, p. 410) terms 'the fragility of things': 'From the perspective of the endurance and quality of life now available to the human estate in its cross-cultural entanglements, in its exchanges with nonhuman force fields, and in the reverberations back and forth between several human and nonhuman processes, we once again inhabit a fragile world'. The drama of the meltdown of the cryosphere is intimately linked to the transience of ice. It disappears without a trace. As Simone Hancox (2013, p. 57) notes in her review of contemporary ecological art engaging with ice, 'in thawing, the ice transforms: from solid to liquid to gas, from order to disorder, from rigidity to movement, from form to formlessness'.

Ice coring or ice-core drilling has, in the last fifty years, transformed the way climate change science works. Because of the ice core record, evidence has mounted that climate change is incredibly rapid, and that the planet has an extremely sophisticated, nonlinear, dynamic climate system. But the information 'trapped' in ancient ice also shows the cumulative effects of the human footprint on the composition of the atmosphere, as well as the functioning and vulnerability of terrestrial and oceanic ecosystems across the globe. Ice cores are an integral part of climate models. Ice allows scientists to travel back in time. Yet, the temporality of ice is intensely precarious. Rising atmospheric and ocean temperatures propel its fragility, its susceptibility to sudden disappearance.

Within human scales of time, the disappearance of ice is often difficult to perceive. As Hancox (2013, p. 54) notes, the temporality of vanishing ice has become 'a cultural meme symbolizing both anthropogenic climate change and an uncertain and threatening future for the planet. Yet for many citizens, this global thaw still appears too distant – both geographically and temporally – to apprehend the actuality of this phenomenon and their own ecological relationship to it'. In the polar regions, however, this disappearance can be detected within the span of one generation.

Ice-sheet modelling thus allows the tracing of ice floes and the movement of glaciers in an attempt to understand glacial cycles and evolution, and future ice-stream dynamics. Modelling ice is a way of modelling temporality: climate modelling of ice dynamics informs contested projections of future geophysical conditions on Earth, where ice is inscribed with time as a mode of anticipating nature. As environmental historians Alessandro Antonello and Mark Carey argue, the practices involved in drilling, analysing, discussing and using ice cores, both for science and broader climate or environmental policies, take part in constituting the temporalities of the global environment; they have helped transform the understanding of what Antonello and Carey refer to as 'Earth time, human time, and future time' (2017, p. 184). In this sense, ice-coring practices are intensely performative in the ways they conjecture about climates past (Earth time), define the Anthropocene (human time), anticipate future climates and speculate about novel ecosystems with no-analogue ecological communities (Williams & Jackson 2007, p. 475).

As I have discussed elsewhere (Salazar 2018), as temporal probes drilling through the very stuff of time, ice cores have brought about their share of surprises, showing the succession of glaciations and the key role of the coupling of the climate with carbon cycle and greenhouse gases. They have also revealed rapid climate instabilities and, in the case of Antarctica, the presence of a huge network of sub-glacial lakes. The weight of ice's materiality, in effect, creates worlds past and future. Jessica O'Reilly (2016, p. 29) argues, in an ethnographic account of the work of glaciologists and climate scientists in Antarctica, that climate science, 'with its prognostic hopes pinned to the global climate model, reifies the global scale into future imaginations'. Antonello (2016, p. 77) argues that Antarctic ice sheets have been understood as entities in time, as simultaneously stable and unstable earthly bodies with their own 'geo-histories and potential futures, with changing narrations over time'. In effect, the Antarctic ice could well play a vital part in identifying and defining the Anthropocene. Drawing on Ingold's (2012) reading of Deleuze and Guattari, it can be argued that ice is always 'matter in movement, in flux, in variation' or what the latter call 'matter-flow' (Deleuze & Guattari 2004, p. 451). Ice comes to matter as the very 'stuff of time' (Salazar 2018, quoting Ingold 2012). As glaciologists and climatologists have been arguing for some time now, ice is inscribed with time. But ice also becomes time in a relentless flux of folding and unfolding, of making and unmaking, of creation and decay, from the depths to the surfaces and back to the depths again. Both the temporalities of ice and its vertical movements through time and material emerge through their recursive interactions with formative and formidable processes such as ablation, striating, melting, cracking, calving and shelving.

Conclusions: an ethics of 'response-ability' for Antarctica in the Anthropocene

Profound ecosystems change in Antarctica is increasingly linked to the shifting geopolitical dynamics of the southern polar region. There is growing recognition that governance of the polar regions is becoming ever more complex and the impact of change in these regions is being felt – not only in climate change, but also in the way we frame globalisation in the first decades of the twenty-first century. As new global and regional political alliances and alignments come to the fore in the circumpolar north and south, as new microbial life forms with potential commercial and health applications are discovered, and as both the Arctic and Antarctic potentially become a new commodity and resources frontier (Dodds & Hemmings 2015), it is imperative to develop interdisciplinary approaches that can open our eyes into how polar worlds are being imagined and put into practice – discursively and materially – by a range of actors across different knowledge practices. The problems and challenges posed by the uncertain nature of the future of the polar regions transcend singular disciplines, and their impact

will be felt far beyond the communities normally focused on science, pol-
icymaking and international relations.

The key analytic with which to think the Anthropocene in Antarctica is
the ice melt now occurring at unprecedented levels. As ice melts archives
into Earth, pasts and futures are lost. This brings about a question of ethics
and the need for a new ethical engagement for Antarctica in the twenty-
first century – a new ethics that could perhaps build on what Donna Haraway (2008) and others have termed a relational ethics of 'response-ability'.
The Anthropocene confers a sense of response-ability – an ability to
respond and to be attentive to the processes of ice melting and the disappearance of ice sheets which, despite glimpses of calving glaciers and
moving ice sheets, always exceed the moment of ethical representation. In
an Antarctic context, where ethics and normative principles arise and are
formalised through the Antarctic Treaty System and its associated protocols
and conventions, I think that – besides looking towards these higher guid-
ing principles and rights that have effectively constructed Antarctica as a
space for science, peace and nature conservation – there must be other
ethical responses that have more to do with entangled subjectivities (human
and other-than-human) and that can open our minds to what learning to
live on a damaged planet means in practice. This entails a 'geo-logic' politics mapped on the basis of relational ethics that could at the same time
provide an alternative paradigm for an Anthropocene Antarctica in the early
decades of the twenty-first century.

References

Adhikari, S & Ivins, ER 2016, 'Climate-driven polar motion: 2003–2015', *Science
Advances*, vol. 2, no. 4, pp. 1–10.
Antonello, A 2016, 'Engaging and narrating the Antarctic Ice Sheet: the history of an
earthly body', *Environmental History*, vol. 22, no. 1, pp. 77–100.
Antonello, A & Carey, M 2017, 'Ice cores and the temporalities of the global environ-
ment', *Environmental Humanities*, vol. 9, no. 2, pp. 181–203.
Bauer, AM & Ellis, EC 2018, 'The Anthropocene divide: obscuring understanding of
socio-environmental change', *Current Anthropology*, vol. 59, no. 2, pp. 209–227.
Chakrabarty, D 2009, 'The climate of history: four theses', *Critical Inquiry*, vol. 35, no.
2, pp. 197–222.
Clark, N 2010, 'Volatile worlds, vulnerable bodies: confronting abrupt climate change',
Theory, Culture & Society, vol. 27, no. 2–3, pp. 31–53.
Clark, N 2014, 'Geo-politics and the disaster of the Anthropocene', *The Sociological
Review*, vol. 62, no. 1, pp. 19–37.
Connolly, WE 2013, 'The "new materialism" and the fragility of things', *Millennium:
Journal of International Studies*, vol. 41, no. 3, pp. 399–412.
Davis, M 2008, 'Living on the ice shelf: humanity's meltdown', viewed 1 March 2019,
<http://www.tomdispatch.com/post/174949>.
Deleuze, G & Guattari, F 2004, *A thousand plateaus: capitalism and schizophrenia*, trans. B
Massumi, Continuum, London and New York.

Dodds, K & Hemmings, AD 2015, Polar oceans: sovereignty and the contestation of territorial and resource rights. *Handbook of ocean resources and management*. Routledge, Abingdon.

Fagan, M 2019, 'On the dangers of an Anthropocene epoch: geological time, political time and post-human politics', *Political Geography*, vol. 70, pp. 55–63.

Glasberg, E 2012, *Antarctica as cultural critique: the gendered politics of scientific exploration and climate change*, Palgrave Macmillan, New York.

Hancox, S 2013, 'The performativity of ice and global ecologies', *Performance Research*, vol. 18, no. 6, pp. 54–63,

Haraway, D 2008, *When species meet*, University of Minnesota Press, Minneapolis.

Haraway, D 2015, 'Anthropocene, Capitalocene, Plantationocene, Chthulucene: making kin', *Environmental Humanities*, vol. 6, no. 1, pp. 159–165.

Hemmings, AD, Dodds, K & Roberts, P 2017, 'Introduction: the politics of Antarctica' in K Dodds, AD Hemmings & P Roberts (eds), *Handbook on the politics of Antarctica*, Edward Elgar, Cheltenham, pp. 1–18.

Hogg, AE & Hilmar Gudmundsson, G 2017, 'Impacts of the Larsen-C Ice Shelf calving event', *Nature Climate Change*, vol. 7, pp. 540–542.

Howe, C 2016, 'Timely', *Theorizing the Contemporary, Fieldsights*, 21 January 2016, viewed 1 March 2019, <https://culanth.org/fieldsights/800-timely>.

Howe, C & Pandian, A 2016, 'Introduction: lexicon for an Anthropocene yet unseen', *Theorizing the Contemporary, Fieldsights*, 21 January 2016, viewed 1 March 2019, <https://culanth.org/fieldsights/introduction-lexicon-for-an-anthropocene-yet-unseen>.

Ingold, T 2012, 'Toward an ecology of materials', *Annual Review of Anthropology*, vol. 41, no. 1, pp. 427–442.

O'Reilly, J 2016, 'Sensing the ice: field science, models, and expert intimacy with knowledge', *Journal of the Royal Anthropological Institute*, vol. 22, no. S1, pp. 27–45.

Paglia, E 2016, 'The northward course of the Anthropocene: transformation, temporality and telecoupling in a time of environmental crisis', unpublished doctoral thesis, KTH Royal Institute of Technology, Sweden.

Robinson, KS 2012, *2312*, Orbit Books, New York.

Salazar, JF 2015, 'Anticipating Antarctica in the 21st century: a view from the social sciences', *ILAIA: Advances in Chilean Antarctic Science*, vol. 2, pp. 36–39.

Salazar, JF 2018, 'Ice cores as temporal probes', *Journal of Contemporary Archaeology*, vol. 5, no. 1, pp. 32–43.

Simonetti, C & Ingold, T 2018, 'Ice and concrete: solid fluids of environmental change', *Journal of Contemporary Archaeology*, vol. 5, no. 1, pp. 19–31.

Sörlin, S 2015, 'Cryo-history: narratives of ice and the emerging Arctic humanities' in B Evengård, J Nymand Larsen & Ø Paasche (eds), *The new Arctic*, Springer, Cham, pp. 327–339.

Steffen, W, Broadgate, W, Deutsch, L, Gaffney, O & Ludwig, C 2015, 'The trajectory of the Anthropocene: the Great Acceleration', *The Anthropocene Review*, vol. 2, no. 1, pp. 81–98.

Stephens, T 2018, 'The Antarctic Treaty System and the Anthropocene', *The Polar Journal*, vol. 8, no. 1, pp. 29–43.

Szerszynski, B 2012, 'The end of the end of nature: the Anthropocene and the fate of the human', *Oxford Literary Review*, vol. 34, no. 2, pp. 165–184.

Voosen, P 2017, '2.7-million-year-old ice opens window on past', *Science*, vol. 357, no. 6352, pp. 630–631.

Waters, CN, Syvitski, JPM, Gałuszka, A, Hancock, GJ, Zalasiewicz, J, Cearreta, A, Grinevald, J, Jeandel, C, McNeill, JR, Summerhayes, C & Barnosky, A 2015, 'Can nuclear weapons fallout mark the beginning of the Anthropocene epoch?', *Bulletin of the Atomic Scientists*, vol. 71, no. 3, pp. 46–57.

Williams, JW and Jackson, ST 2007, 'Novel climates, no-analog communities, and ecological surprises', *Frontiers in Ecology and the Environment*, vol. 5, no. 9, pp. 475–482.

Wolff, EW 2014, 'Ice sheets and the Anthropocene' *Geological Society, London, Special Publications*, vol. 395, no. 1, pp. 255–263.

Yusoff, K 2013, 'Geologic life: prehistory, climate, futures in the Anthropocene', *Environment and Planning D: Society and Space*, vol. 31, no. 5, pp. 779–795.

Yusoff, K 2018, *A billion black Anthropocenes or none*, University of Minnesota Press, Minneapolis.

Zalasiewicz, J 2008, *The Earth after us*, Oxford University Press, Oxford.

Zarankin, A & Salerno, MA 2014, 'The "wild" continent? Some discussions on the Anthropocene in Antarctica', *Journal of Contemporary Archaeology*, vol. 1, no. 1, pp. 114–118.

Part 2
Cultural texts and representations

6 Ice and the ecothriller

Popular representations of Antarctica in the Anthropocene

Elizabeth Leane

In *The Great Derangement: Climate Change and the Unthinkable*, Amitav Ghosh (2016, p. 8) identifies a '[broad] imaginative and cultural failure that lies at the heart of the climate crisis'. Ghosh is concerned specifically with 'serious' or 'literary' fiction, which, he argues, foregrounds the everyday over the improbable, exceptional or catastrophic event. Novels featuring such events, he observes, are automatically banished to the 'generic outhouses' of fantasy, horror and science fiction (p. 24). Ghosh does not mention one of the biggest selling of all popular genres: the thriller. However, this genre – particularly in its geopolitical form – trades on the improbable, with the hero routinely performing spectacular feats of daring to prevent imminent nuclear war, global epidemic outbreak or, more recently, environmental catastrophe.

Ghosh is not alone in overlooking the thriller: literary critics, even those interested in genre fiction, also largely ignore it. In her introduction to *Popular Fiction and Spatiality*, Lisa Fletcher (2016, p. 4) notes that this popular genre has produced less criticism than any other. This is even more true of the thriller's environmentally inflected subgenre, the ecothriller. While the term (sometimes hyphenated or separated into two words) is current in popular discourse – a Google search produces more than a hundred thousand hits – it barely registers in academia. Only two of more than two million entries in the Modern Languages Association bibliography mention 'eco(-)thriller'; none includes the term 'environmental thriller'; only nine can be found from a general search on the combined terms 'thriller' and 'environment(al)'. Richard Kerridge's short article 'Ecothrillers: environmental cliffhangers', published in 2000, remains one of very few critical efforts to directly address the subgenre as a whole. And yet, the thriller, with its tendency to operate across large, often global, spaces, its willingness to deal with the improbable and its emphasis on action, is a particularly interesting genre to consider in light of Ghosh's observations about the failure of imaginative literature to address the challenges of the Anthropocene.

In this chapter, I therefore examine the relationship between setting, plot and character in a group of thrillers – many of them ecothrillers – that take place in a specific natural environment: the Antarctic icescape. As several other chapters in this volume demonstrate (see particularly those by Salazar, Hemmings, Nielsen and McGee), ice has taken on new cultural salience in the age of the Anthropocene. Even as scientists analyse the gas contained in ice cores in order to predict the future of our planet, warm currents undermine Antarctic glaciers, threatening future icesheet collapse and rising sea levels. Ice's increasing prominence in the social, media and political spheres has brought the attention of critics to its distinct physical properties, as well as to the diverse set of functions and meanings it holds for human communities.[1] Historian Sverker Sörlin, arguing for the recognition of the present as a 'cryo-historical moment', writes that, 'Ice is a probing element, in which civilization is put to the test, and crisis is a core element in narratives of ice, in recent decades reinforced by projections of climate change science' (2015, p. 327). Long relegated to the margins of cultural consciousness, icescapes are currently moving rapidly towards its centre.

Consumed by millions of readers, thrillers offer a revealing insight into how popular culture is currently imagining human relationships with Antarctica and its ice. In an earlier analysis (Leane 2016, pp. 33–37), I briefly examine how the unreliable and unreadable icescapes of Antarctic thrillers can be read as a material metaphor for the large-scale conspiracy at the heart of the thriller narrative. At the same time, I suggest, polar icescapes function as actors in the narrative, forming ambiguous alliances with both the human heroes and villains. Here, I expand on this reading to consider the way in which the icescape, as both metaphor and actor, works to mediate the tension between the local and the global that characterises both environmental and geopolitical thrillers. I will conclude by turning from spatiality to temporality, suggesting that the urgent timeline of the thriller narrative, extrapolated onto real-world scenarios, might be constraining popular understandings of Antarctica's future. The chapter thus aims to provide insight not only into popular perceptions of Antarctica in the early twenty-first century but also into the questions of scale that are integral to the literature of the Anthropocene.[2]

The rise of the Antarctic (eco)thriller

Popular fiction has dominated imaginative written responses to the Antarctic region, forming a barometer of knowledge and perception of the far south, while also reflecting developments in the mass-market novel. The late nineteenth and early twentieth centuries, for example, saw a stream of Antarctic 'lost world' novels, which dovetailed into a growing interest in the continent by pulp science fiction writers in the 1920s and 1930s. Their speculative works oscillated between techno-optimistic terraforming tales and cautionary stories of hubristic scientists punished in horrible ways, usually by defrosted aliens. Throughout these periods, the polar imperial adventure tale maintained a fairly steady presence, initially directed at a juvenile readership, then transformed after the Second World War into the adult thriller. Antarctic science fiction, dystopias, horrors and even romance novels also appeared but, since the mid-century, genre writers and readers have engaged with Antarctica primarily through the thriller.

Using online bibliographical resources, it is possible to identify dozens of Antarctic thrillers published over the last seventy years – the number depends on how tightly the genre is defined, how tightly 'the Antarctic' is defined and how central to the story world the region needs to be to qualify.[3] Historically, the first adult Antarctic thriller is probably British novelist Hammond Innes's *The White South* (1949), an international bestseller that was subsequently made into a film featuring American Western star, Alan Ladd. Innes returned to the Antarctic setting late in his career, with *Isvik* (1991) and *Target Antarctica* (1993). By this stage, he had been joined by many other authors, with bestselling writers such as Clive Cussler, Matthew Reilly and Michael Crichton publishing thrillers set wholly or partly on Antarctica during the 1990s and early 2000s.[4]

Like other popular novels, thrillers are often dismissed as 'airport reading', unworthy of serious discussion, but their potential to both reflect and affect readers' worldviews is considerable. Critical geopolitics scholar Klaus Dodds raises questions about the impact of thriller narratives and character stereotypes on the way people understand the contemporary world:

> How does the action-thriller reinforce or unsettle particular framings of [political events]? Do these artistic interventions help to constitute public understandings of key actors and places, and are they more significant when watched and engaged with by audiences that are not likely to have any experience of the places cited? … are there different kinds of geopolitics based on action-thrillers, dramas, horrors, disasters, romance and fantasy?
>
> (Dodds 2014, p. 121)

For Dodds, thrillers (and popular texts more generally) do not – or do not only – *represent* geopolitics, but also *produce* a particular geopolitics, when their narrative conventions are applied by political actors to reveal events. And while Dodds, like many critical geopolitics scholars who incorporate analysis of popular culture into their work, concentrates on thriller films, thriller novels have by no means lost traction with readership. Dan Brown's *The Da Vinci Code* (2003), for example, is among the highest-selling novels of all time. And, as this example suggests, thriller novels form the basis of many of the most successful thriller films. Given that Antarctica is a place that only a minute percentage of humanity ever experiences directly, Antarctic thrillers might well have a disproportionate influence on the way people understand the continent.

While the term 'thriller' suggests a homogenous group, this genre has, like any other, considerable internal variation. In his analysis of the American thriller, Paul Cobley (2000, p. 3, emphasis in original) begins from 'the premise that the notion of the conspiracy is so wide and accommodating that it *enables* an expansive range of diverse texts'. The genre thus divides into numerous subgenres, some of which are better suited to an Antarctic setting than others. While it is possible to find a range of thriller subgenres set on the continent,[5] the remote, isolated and dangerous icescape lends itself more readily to the action thriller, the technothriller and the ecothriller – genres which often overlap. In addition to the possibilities for danger inherent in the extreme environment, the geopolitics of Antarctica offers an unexpectedly rich resource for thriller writers. Matthew Reilly, author of the bestselling *Ice Station* (1997), has observed that a continent which is unowned and largely unexploited, but remains the subject of a series of contested national claims, provides ample scope for secret government machinations (Leane 2012, p. 23). Intrigue usually revolves around mineral resources, espionage activity, secret military installations, covert territorial manoeuvres and corporate greed. Thus, the Antarctic thriller frequently involves large-scale – often global – conspiracies.

Australian novelist LA Larkin's *Thirst* (2012) is a good example. Set in the near future, the plot revolves around the Chinese CEO of a global private equity firm who is intent on using explosives to blast off part of an Antarctic glacier to sell ice to a water-short China – an action which could set off a catastrophic series of ice sheet collapses. It transpires that the CEO is being manipulated by his father, a Chinese military general, who wants to access Antarctica's minerals – specifically, rare earth elements used in weapons construction. The hero is an Australian glaciologist who, once he has defeated the villains, goes on to become the face of a global campaign to address climate change. While the action is largely set near the rapidly retreating Pine Island Glacier in the unclaimed part of Antarctica known as Marie Byrd Land, it also incorporates episodes in Australia and China, as well as the involvement of a Norwegian-led tourist expedition on a Russian icebreaker.

As this summary suggests, *Thirst* can be straightforwardly classed as an ecothriller (and its author is a former climate change consultant). If, as seems to be the case in popular usage, this subgenre is defined as a thriller in which environmental exploitation is a central part of the narrative, then Antarctic fiction is marked by a preponderance of ecothrillers, although some early titles pre-date the term itself. This is not to say that all of these texts can be considered pro-environmental. 'Ecothriller' as popularly used is a loose term applied thematically with little regard to the text's environmental politics (which are not always themselves straightforwardly categorisable). For example, the term is regularly used to describe Michael Crichton's controversial *State of Fear* (2004), which is partly set in Antarctica, although the villains of this novel are environmental activists who manufacture apparent climate-change disasters. However, a number of Antarctic novels (like *Thirst*) can be classed as ecothrillers in the stronger sense that the central threat is directed at the natural environment. In this context, the burgeoning Antarctic thriller genre of the 1980s and 1990s can be read not – or not only – as an artefact of an overall rise in thriller publication during this period (Anderson 2007, p. 69), but also as a response to the fierce debates about Antarctic mining that arose prior to the (eventually stalled) development of a minerals convention in 1988 and its subsequent replacement by an environmental protection protocol in 1991.

Since the turn of the twenty-first century, climate change has become a common issue in Antarctic thrillers, but it is rarely the specific source of threat. As Adam Trexler (2015, p. 14) has observed with respect to 'chiller' fiction, 'condensing the distributed impersonal causes of global warming on to a climate villain' is difficult. Thus, material resources – including, increasingly, bioprospecting and ice itself – are more often at the heart of the plot, even as the characters operate within a climate-conscious environment and are sometimes themselves climate scientists.

Whether or not they engage with environmental politics, all Antarctic thrillers feature the physical environment – the icescape – as an integral part of their action. As I have observed elsewhere (Leane 2016, p. 31), 'ice' is the word most frequently featured in their titles: examples include *Beneath the Ice, Beneath the Dark Ice, Out of the Ice, A Grue of Ice, The Icemen, Ice Station, Ice Reich, Ice Wolf, IceFire, Black Ice, Red Ice* and *The Ice*. [6] This is reinforced by the novels' covers, which very often feature an icy scene, usually combined with a human figure or a vehicle of some sort to indicate the pace so important to the genre. Given that both titles and covers are paratextual features that are as much about marketing as narrative content, we might conclude that the Antarctic location is simply an exotic draw for the reader. However, as I will argue in the next sections, the icescape's role in these thrillers can be far more integral.

Global plot, local action

At the heart of many thriller novels lies a tension between the local and the global: the conspiracy against which the hero must fight is often global in scope, but the action through which it is resolved is usually highly localised, in the form of individual combat. This disparity can be symbolically resolved by locating the scene of combat in a place that physically evokes the opacity, complexity and mysteriousness of the conspiracy. In an urban thriller this might be labyrinthine city streets. In the ecothriller it is frequently a natural setting: the subterranean darkness of a cave system, the dense undergrowth of a jungle or indeed the crevasses of an Antarctic icescape.[7]

A central function of the icescape in the Antarctic thriller, then, is to mediate between the disparate scales at work in the plot. The global scale on which the conspiracy functions contrasts markedly with the action of the narrative, which almost always involves an individual hero and a villain, who at some point take part in combat. While this combat might be remote and technologically mediated, and sidekicks of varying prominence might take part on both sides, the vulnerability of the hero's body is a central element in the narrative. His or (less often) her body must be located in a particular place, which is ideally a dangerous and threatening one that heightens the sense of vulnerability and isolation.[8] But the icescape does more than this: it also functions analogically, mimicking the human-generated conspiracy plot that the hero must negotiate. Hard and solid, but liable to break, collapse or melt at the wrong moment, ice is unreliable and unpredictable. Antarctic conditions mean that visibility is often low, due to fog, blizzard or the prolonged winter night. Together, these elements act as an excellent metaphor for a labyrinthine plot in which motives are concealed, allegiances are ambiguous and apparent certainties can be undermined in an instant.[9] The icescape enables a materialisation of a conspiracy that otherwise remains unreadable and opaque, allowing local action and global plot to symbolically meet.

A similar 'downloading' of the global onto the local occurs with the human characters. The disparity between global conspiracy and local combat means that large-scale forces and issues are compressed into individual bodies. Several critics have noted this dynamic in relation to crime fiction generally and the thriller more specifically.[10] David Schmid (2012, p. 19) points to crime fiction's tendency to 'provide accurate diagnoses' of social problems but to be 'much less forthcoming about solutions that are anything except individual'. This becomes more evident, he notes, 'when the genre deals with units of space larger than the city, up to and including the globe' (p. 19). Kerridge (2000, p. 249) identifies the same tendency as a problem with ecothrillers – they condense 'large diffuse issues into short tense dramas between a few individuals', so that 'ecological solutions are replaced by quick, violent fixes'. In the Antarctic thriller scales are regularly confused in this way, with planetary-level, systemic problems given local, personal resolutions.

A useful exemplar of the projection of global problems onto local action process is Crichton's *State of Fear*. As noted above, the plot of this controversial bestseller sees environmental activists manufacturing a series of climate-related disasters, including the deliberate blasting off of the Ross Ice Shelf, in an attempt to manufacture public concern about climate change. Following its publication, Crichton was famously invited by a Republican senator to testify on climate change issues before a US Senate Committee and also met with President George W. Bush in the White House (White 2011, p. 18). In an 'Author's Message' at the end of the book, Crichton (2005, p. 676) writes that his novel, 'in which so many divergent views are expressed, may lead the reader to wonder where, exactly, the author stands on these issues'. Here, by implying that the novel's politics can be looked for in the explicit yet varied views held by its characters, Crichton disingenuously overlooks the impact of the action: what happens to characters who hold particular views. The narrative certainly provides space for the celebrity environmentalist Ted Bradley to express his climate change activism, but he is punished by being hacked to death by cannibalistic Pacific Islanders. The everyman-hero Peter Evans is also allowed to air his increasingly sceptical views of climate change science, and is rewarded with survival, a job offer and a relationship with an 'extremely beautiful woman' (2005, p. 72). Although the conspiracy at the thriller's centre is shadowy and dispersed and the threat – the inundation of California by a tidal wave – large-scale, the victory and punishment are worked out on a local, indeed bodily, level, leaving little doubt about how the author stands on the issues.[11]

While few Antarctic ecothrillers share the politics evident in *State of Fear*, the tendency to combat planetary threats with heroic (and usually masculine) individual action is certainly characteristic of these texts. A tagline of *Thirst* asks, 'Can one man stop global disaster?'. In the narrative, multinational capitalist greed and high-level military megalomania are condensed onto the villain and his father, and altruistic science onto the hero and his female sidekick/love interest. In the end, global disaster is only partially averted – although the general's mining plans are exposed, the icesheet is exploded off and the climate catastrophe worsened. However, the hero physically defeats the villain, so positive closure is achieved at the primary level of the individual.

The process of condensing global forces onto human characters is complicated, however, when the icescape itself becomes part of the action. Its constant movements (both slow and sudden) and its ability to transform states gives ice a unique sense of animate existence. In a number of the thrillers I have examined, the ice advances the plot, as human and nonhuman actors become entangled (Leane 2016, pp. 33–37). In the following section, I expand on the various roles the ice takes on in the Antarctic thriller, and the possibilities they close down or open up for ecocritical readings.

Ice as nonhuman actor in the ecothriller

The weakest sense in which the ice functions as a nonhuman character in the thriller is when it is constructed as a victim – weakest, because this leaves the ice passive, an idealised and vulnerable environment present only to be exploited and saved by humans. The tagline to one Antarctic thriller, *Cold War*, reads 'Unspoiled, uninhabited, under attack …!' (Preisler 2001). The continent is, indeed, under attack in these texts: its (real and imagined) resources – gold, uranium, caesium, oil – are plundered by greedy corporations or sinister nations; it is used as a testing ground for nuclear weapons; its ice shelves are blasted off for various reasons; its extremophile organisms are put to nefarious purposes. The hero might explicitly cast him or herself as the champion or saviour of the continent. In *Thirst*, Luke Searle sees his efforts to stop the blasting off of the ice as an extension of his scientific work, asking himself at one particularly desperate moment whether he has 'lost the battle to save' the glacier (Larkin 2012, p. 326).

In a number of Antarctic thrillers – particularly technothrillers – this victimisation stretches to forcing the ice to act as a weapon, with villains taking advantage of its huge extent to ratchet up their destructive impact to planetary level. In Judith and Garfield Reeves-Stevens' *IceFire* (1998), for example, the villain – another Chinese nationalist – explodes a thermonuclear device above the Ross Ice Shelf, its sudden thrust into the ocean calculated to create an enormous tsunami destined for the Californian coast.[12] Other texts cast the ice in a more actively hostile role, as a nonhuman enemy that the hero must fight alongside the villain. Innes's novels in particular tend to be interpreted this way, with the nonhuman environment given higher billing by critics than the human villain. According to his entry in *Twentieth-Century Crime and Mystery Writers*, Innes writes of 'the conflict of man against man, good against evil, but this is no more than an undercurrent to his main theme, that of man against the overpowering force of implacable nature' (Hughes 1991, p. 589).

However, these duelling binaries – 'man against nature', 'man against man' – are too simplistic to describe the action of these thrillers. In an examination of Clive and Dirk Cussler's climate change ecothriller *Arctic Drift* (2008) – one of very few academic treatments of Cussler's bestselling novels – Adam Trexler (2012, pp. 305, 302) points to the way in which 'things exert their agency in conjunction with characters' in the novel, arguing that this exemplifies a tendency of nonhuman agency to be much more clearly presented in genre fiction than literary fiction. Although Trexler does not discuss the way ice operates in the narrative, his observation transfers well to the Antarctic thrillers discussed here. The relationship between hero, villain and nonhuman environment in the thriller is best described not as two sets of binaries, but as a triad in which the actors enter into complex and shifting alliances and antagonisms. In *The White South* (1949), the villain attempts to use ice as a weapon, orchestrating the shipwreck of the hero and his companions in the path of a series of oncoming bergs. But in the end, the hero effectively befriends the icebergs bearing down upon him and his marooned companions, boarding one berg and setting up camp upon on it. By the time he has to leave it again, to pursue the villain, the iceberg has 'assumed the friendliness of a home' (Innes 1949, p. 225). In *A Grue of Ice* (first published in 1962), which centres around an elusive Antarctic island, the hero refutes another character's attempt to attach malevolence to the place:

> 'Some unknown rock sticking out into an icy ocean suddenly becomes a killer and the thing that dominates all our lives! I'm frightened. Something wicked and enormous is building up. Even the sky is ghastly. Look, it's getting quite dark – '
> 'Your father is building up the evil', I said. 'He has got to be stopped'.
> (Jenkins 2009, p. 124)

Rather than blaming the icescape, the oceanographer-hero uses his scientific knowledge to enlist its help in defeating the human villain. The latter's demise occurs when an entire glacier collapses on him – an action started by a warm current coincidentally arriving at exactly the right time, but accelerated by the hero. Recognising the natural dynamic at work, he contributes to the collapse with a series of torpedoes. It is the hero's ability to work with, as well as for, a nonhuman agent that enables his, or rather their, victory.

This recognition and deployment of the environment's own agency, latent in the structure of early Antarctic thrillers such as *A Grue of Ice*, becomes more obvious as the decades move on and ecothrillers become more common. By the twenty-first century, the environment's role in the action becomes overt. Again, Larkin's *Thirst* provides a useful recent example. The glaciologist-hero sees the Antarctic as a homely, familiar place, at one point patting the threatened glacier 'as if comforting a sick animal' (Larkin 2012, p. 300). Before long, his adeptness in the extreme environment comes to look like the environment's preference for him: 'there was something peculiar about Searle's connection to Antarctica', reflects the frustrated villain, 'It was almost as if this wretched heaving mass of ice was keeping its protector alive. Ridiculous, of course' (p. 308). But it isn't ridiculous: with his exploitative attitude, the villain is unable to enrol the icescape into his plans. Thus, while the novel explicitly casts the glaciologist as the continent's human saviour, there is a sense in which the ice takes an active part in the battle. In the end, while the hero ensures the villain's death, he is unable to stop the explosive fracturing of the glacier. Searle himself only survives by clinging to what has become a berg; in a sense, the ice *saves* him. The final chapter sees him in self-imposed exile from his beloved ice, now a celebrity heading up a foundation for climate-change refugees, and travelling the world to raise awareness, despite his aversion to media appearances. As I have suggested above, this conclusion could be read critically: so long as the hero survives and is personally rewarded, the compulsory happy ending is achieved, the collapsing ice and warming climate notwithstanding. On the other hand, it could be read more positively (in an environmental sense) as a refusal to use individual combat as a narrative solution to a global problem: the hero might have won the physical fight, but he is unable to stop the environmental disaster in this way, and must instead work on a global stage, through advocacy and consciousness raising. The answer to the tagline, 'Can One Man Stop Global Disaster?', turns out to be no – at least, not through local hand-to-hand combat and physical heroics, but possibly through global, mediated activism.

Can the ice have a more active role in the Antarctic thriller than as a weapon brandished by the villain, the hero's 'sidekick' or a victim the hero is trying to save? If the narrative agency of ice in the Antarctic thriller were placed on a continuum, one end would indicate complete passivity (the ice has no active role in the narrative) and the other wholesale activity (the ice ultimately usurps the role of protagonist or antagonist). Formally, is difficult to imagine this latter end of the continuum. The thriller is usually told from the hero's point of view: a shift to the ice-as-hero would thus require the narrative to be focalised through the nonhuman. This stylistic innovation would be difficult to reconcile with the need for a fast-paced plot and would presumably move the novel out of the thriller (and indeed the genre fiction) category and into the literary register.

An arguable exception is James Follett's novel *Ice* (1978), in which an enormous tabular iceberg breaks off Antarctica and makes its way north through the Atlantic. As the berg has sheared off a mountain top that remains trapped within its ice, increasing its density, it does not float but lurks unseen in the ocean, its top just under the surface, sinking ships and breaking undersea cables. The action occasionally shifts away from the human characters and to give brief descriptions of the iceberg's onward movement, which is highly anthropomorphised and expressed in language usually reserved for human villains:

> The ice moved.
>   It had proved itself. It was unstoppable. Indestructible. It moved very slowly. Almost as if it was possessed of a blind but certain instinct.
>   And that instinct was to kill.
>
> (Follett 2004, pp. 177–178)

The berg here is both indifferent ('blind') and yet somehow sentient, having if not intention, then at least 'instinct'. Humanity, however, overlooks the nonhuman in its rush to attribute agency. The West thinks the series of destructive maritime incidents are caused by the Soviets, the Soviets think they are caused by the West, and the world sits on the brink of war. It is only when the enemy is finally recognised as natural, not human, that action can be taken. The berg – now approaching Manhattan – is bombarded with nuclear weapons and, after the failure of this assault, towed in the opposite direction by a huge flotilla of vessels, with ships from 'virtually every nation in the world' cooperating in the rescue effort (p. 191). This, then, might be considered a thriller in which ice is the central villain. However, with the revelation to the human characters of the enemy's identity as natural, the thriller merges into a neighbouring genre, the disaster narrative.

The icescape, then, can function in complex ways in Antarctic ecothrillers, both as a material metaphor for the large-scale conspiracy at the heart of the plot, and an important nonhuman actor within that plot. Sometimes the ice is reduced to victim, weapon or enemy; at other times, villain, hero and icescape become entangled in the action and the outcome depends on human–nonhuman alliances as much as human-to-human combat. These thrillers provide what Ghosh (2016, p. 64) describes, in a very different context, as 'techniques of storytelling [in which] nonhumans … create the resolutions that allow the narrative to move forward'. This reading might be applied to thrillers set in many remote natural environments. Certainly, Ralph Crane and Lisa Fletcher (2016, p. 20) make some similar observations in their analysis of subterranean thrillers, noting the tendency for the hero to initially misrecognise the natural environment as an enemy, before realising that this role is actually occupied by a human. However, as I have observed above, ice has a particular salience at the present moment because of its role in preserving climates past (in the gas bubbles in ice), as well as its disconcerting tendency to crack, shift, turn and melt, thereby reminding us of our planet's unpredictability, changeability and liveliness. If, as Ghosh (2016, p. 33) implies, our culture needs to come to terms with the 'uncanny intimacy of our relationship with the nonhuman', then imaginative stories set in icescapes should be at the forefront of our attention.

Deadlines, countdowns and the future of Antarctica

I have been focusing thus far on the way in which space works in the Antarctica thriller: the tendency for a global threat to be collapsed onto and worked out through local characters and the way the icescape functions in this process. However, thrillers rely not only on constant action occurring in hostile places, but also on a sense of temporal urgency. Their narratives are, in essence, suspenseful and fast paced, which means that the plot inevitably involves a deadline. In Larkin's *Thirst*, as in many genre thrillers, this is explicit: each chapter is headed 'T minus…' a certain number of days and minutes (in the style of a NASA countdown) – with 'T' indicating the time until the ice shelf is exploded. (Readers are not aware of the exact nature of this impending cataclysm until some way into the book.) Just as the spatiality of the thriller – its inability to address a global threat except in terms of local action – presents a problem for dealing with the vast scale of the climate change and other challenges of the Anthropocene, so the finite timeline of the thriller – the countdown to midnight – raises issues for the genre's ability to deal with the extended temporal scale of these planetary challenges.

The increasing anxiety attached to Antarctica in the twenty-first century – around tourism and other human interactions; around geopolitics and resources; and especially around the impact of climate change and resulting sea-level rise – has, I suggest, generated a kind of real-world thriller narrative in which humans see themselves in a race against time. Yet, climate change and many of the other environmental challenges facing the continent lack a firm deadline to match the urgency of the issues. This narrative problem is often solved in public discourse by a focus on the year 2048, the date at which, numerous media articles suggest, the Antarctic Treaty, or at least the associated environmental protection protocol, 'expires' or requires renewal.[13] Already under threat from a warming climate, this argument goes, the continent by mid-century could be subject to a geopolitical free-for-all as nations vie for its mineral resources, which will in turn be rendered more accessible by the melting ice. In actuality, neither the Treaty nor any of its instruments (including the environmental protocol and its mining ban) has an expiry date and this dramatic scenario has been convincingly debunked as a widespread myth (Gilbert & Hemmings 2015, pp. 29, 31). However, in the context of the thriller narrative, it is easy to see why this date has such resonance: we must save Antarctica, and there is a countdown to 2048.[14] This is, I would argue, an illustration of Dodds's idea of a 'geopolitics based on action-thrillers', with fiction providing a narrative template that enables the mistaken idea of a countdown to a precise date to be readily perpetuated. The problem then becomes not so much (to return to Ghosh's ideas) fiction's inability to deal with the catastrophic event, but rather its tendency to project the thriller's focus on the time-bound event onto a planet with no specific deadlines.[15]

A corollary of adopting the thriller template as a way to understand environmental crisis is that heroic scientific efforts – such as the 'race' to find a million-year ice core (discussed in Alan Hemmings's chapter in this book)[16] – become particularly attractive as a symbolic counter to this looming threat. Yet, this idea constrains our imagination as well as enabling it. It suggests that our problems are best solved through heroic individual achievement, rather than collective mundane action, such as reducing the greenhouse gas emissions which are causing climate change. And, coming as part and parcel of the narrative elements of the thriller, it can limit the way in which we imaginatively plot the future.

It is unlikely that thriller narratives are going to lose their glamour through the efforts of literary critics. Nor should we necessarily want them to: as I argue above in regard to Antarctic ecothrillers, they can offer narratives of human/nonhuman alliance that reflect new ways of thinking in the Anthropocene. I suggest, however, that we also remain alert to what their narrative conventions distract us from. Rather than anticipating dramatic deadlines, we must attend to the challenges of the Anthropocene that are here, now, ongoing and cumulative. Rather than focusing attention on villains and heroes who are exceptional and polarised, we should watch for those that are collective, multiple, overlapping and ordinary.

Notes

1 See for example Dodds 2018; Antonello 2017; and Bjørst 2010.
2 For a fascinating discussion of scale and contemporary literature, see Clark (2015), particularly ch. 4 and ch. 5. Clark focuses primarily on literary rather than popular texts, but his observations (pp. 75–78) of ecocritical literary scholars' problematic tendency to treat individual characters as miniaturised versions of large-scale social problems is relevant to my argument here.
3 See Leane 2016, p. 29, for a list of several dozen Antarctic-set thrillers – a list that would be longer if compiled now.
4 Cussler, who primarily writes maritime adventure thrillers, has set several novels partly or mostly in the Antarctic region, including *Shock Wave* (1996), *Atlantis Found* (1999) and (with Jack du Brul) *The Silent Sea* (2010). See also Reilly 1998; and Crichton 2005.
5 David Mariner's *Symbol of Vengeance* (1975), for example, could be classed as a psychological thriller.
6 A word-cloud based on three dozen Antarctic thriller titles can be found in Leane 2016, p. 31: 'ice' is by far the dominant term, followed by 'white'. *A Grue of Ice, Ice Station* and *IceFire* are discussed below. For details of the other novels, see the References.
7 For example, in an analysis of Jeff Parker and Steve Lieber's thriller *Underground*, Ralph Crane and Lisa Fletcher (2016, pp. 9, 11) read 'the heroine's blind descent into the uncharted depths of a limestone cave system' as 'an analogy for her deep investigation of a labyrinthine conspiracy involving big industry and organized crime'.
8 The very evidently gendered nature of both authors and heroes of Antarctic thrillers is beyond the scope of this chapter, but this might be gradually changing. Recent female authors of Antarctic thrillers include Sarah Andrews, Karen Dionne, LA Larkin and Ann Turner. Female heroes are also becoming more common, including in thrillers authored by men. James M Tabor's *Frozen Solid* (2013) and Matt Dickinson's *Black Ice* (2002) are examples.
9 This is distinct from the 'pathetic fallacy' in which the environment simply reflects the protagonist's emotions.
10 There is no critical consensus on whether the thriller is a subset of crime fiction, but it seems clear at least that these are closely related genres which overlap in places.
11 The network of villains and heroes in *State of Fear* is complicated. The central villain is Nicholas Drake, the director of the National Environmental Resource Fund, but Bradley functions as a secondary villain, spying on Drake's enemies. Similarly, the novel is focalised through Evans, but the scientist-cum-secret-agent John Kenner whom he joins with also displays many of the traditional qualities of a thriller hero. The point here is that the conflict between large-scale forces is worked out bodily through their representatives.
12 The idea of manipulating the natural environment to create planetary-scale weapons has its roots in the Cold War and can be found in Antarctic fiction of that era. See Leane [forthcoming].
13 See for example Lockyer 2015; McKie 2017; and Marshall 2009. These are but a few of the numerous examples in the mainstream media.
14 The factual core in the mythologisation of 2048 lies in a clause in the Protocol on Environmental Protection to the Antarctic Treaty stating that from this date (fifty years after its ratification) any party can call for a review conference at which amendments can be made under certain conditions. However, as Gilbert and Hemmings (2015, p. 29) point out, the Protocol can be modified at present through consensus agreement, and the Antarctic Treaty itself reached a similar deadline – i.e. one after which any party can call for a review – in 1991, without dramatic (or indeed any) consequences.

15 There are connections here to arguments against the notion of a distinct 'tipping point'; see for example Clark 2015, pp. 80, 91.

16 For the framing of this scientific project as a 'race', see Denholm 2016.

References

Anderson, P 2007, *The triumph of the thriller: how cops, crooks, and cannibals captured popular fiction*, Random House, New York.

Antonello, A 2017, 'Engaging and narrating the Antarctic ice sheet: a history of an earthly body', *Environmental History*, vol. 22, pp. 77–100.

Barker, N & Masters, A 1986, *Red ice*, Constable, London.

Beck, G 2009, *Beneath the dark ice*, Pan Macmillan, Sydney.

Bjørst, LR 2010, 'The tip of the iceberg: ice as a non-human actor in the climate change debate', *Études/Inuit/Studies*, vol. 34, no. 1, pp. 133–150.

Charbonneau, L 1991, *The ice*, Pocket-Simon & Schuster, New York.

Clark, T 2015, *Ecocriticism on the edge: the Anthropocene as a threshold concept*, Bloomsbury Academic, London and New York.

Cobley, P 2000, *The American thriller: generic innovation and social change in the 1970s*, Palgrave Macmillan, Houndsmills.

Crane, R & Fletcher, L 2016, 'Cave genres/genre caves: reading the subterranean thriller' in L Fletcher (ed.), *Popular fiction and spatiality: reading genre settings*, Palgrave Macmillan, London, pp. 9–24.

Crichton, M 2005 [first published 2004], *State of fear*, HarperCollins, London.

Cussler, C 1996, *Shock wave*, Simon & Schuster, New York.

Cussler, C 1999, *Atlantis found*, GP Putnam's Sons, New York.

Cussler, C & du Brul, J 2010, *The silent sea*, GP Putnam's Sons, New York.

Denholm, M 2016, 'Antarctic research race on for million-year-old ice extraction', *Australian*, 11 December 2016, viewed 5 March 2019, <https://www.theaustralian.com.au/national-affairs/climate/antarctica-research-race-on-for-millionyearold-ice-extraction/news-story/a0018ecc98077b6d22376e64551bb6a9>.

Dickinson, M 2002, *Black ice*, Hutchinson, London.

Dietrich, W 1998, *Ice Reich*, Warner, New York.

Dodds, K 2014, *Geopolitics: a very short introduction*, 2nd edn, Oxford University Press, Oxford.

Dodds, K 2018, *Ice: nature and culture*, Reaktion, London.

Fletcher, L 2016, 'Introduction: space, place, and popular fiction' in L Fletcher (ed.), *Popular fiction and spatiality*, Palgrave Macmillan, London, pp. 1–8.

Follett, J 2004 [first published 1978], *Ice, U700, Churchill's Gold*, Arrow-Random House, London.

Ghosh, A 2016, *The great derangement: climate change and the unthinkable*, University of Chicago Press, Chicago and London.

Gilbert, N & Hemmings, AD 2015, 'Antarctic mythbusting', *Antarctic*, vol. 33, no. 3, pp. 29–31.

Henrick, H 1994, *Ice wolf*, HarperCollins, New York.

Hughes, DB 1991, 'Innes, (Ralph)' in L Henderson (ed.), *Twentieth-century crime and mystery writers*, St James Press, Chicago and London, pp. 588–589.

Innes, H 1949, *The White South*, Harper & Brothers, New York.

Innes, H 1991, *Isvik*, Chapmans, London.

Innes, H 1993, *Target Antarctica*, Chapmans, London.

Jenkins, G 2009 [first published 1978], *A grue of ice*, Authors Choice Press-iUniverse, New York and Bloomington.

Kerridge, R 2000, 'Ecothrillers: environmental cliffhangers' in L Coupe (ed.), *The green studies reader: from romanticism to ecocriticism*, Routledge, London and New York, pp. 242–249.

Larkin, LA 2012, *Thirst*, Pier 9-Murdoch Books, Sydney.

Leane, E 2012, *Antarctica in fiction: imaginative narratives of the Far South*, Cambridge University Press, Cambridge.

Leane, E 2016, 'Unstable places and generic spaces: thrillers set in Antarctica' in L Fletcher (ed.), *Popular fiction and spatiality: reading genre settings*, Palgrave Macmillan, London, pp. 25–43.

Leane, E [forthcoming], 'The Coldest War: imagining geopolitics from the bottom of the Earth' in A Hammond (ed.), *The Palgrave handbook to Cold War literature*, Palgrave Macmillan, London.

Lockyer, A 2015, 'A Cold War on Australia's doorstep', *HuffPost Australia*, 28 August 2015 (updated 15 July 2016), viewed 4 March 2019, <https://www.huffingtonpost.com.au/adam-lockyer/a-cold-war-on-australias-doorstep_b_8046812.html>.

McKie, R 2017, 'How Sea Shepherd lost battle against Japan's whale hunters in Antarctica', *Observer*, 24 December 2017, viewed 5 March 2019, <https://www.theguardian.com/environment/2017/dec/23/sea-shepherd-loses-antarctic-battle-japan-whale-hunters>.

Mariner, D 1975, *Symbol of vengeance*, Robert Hale, London.

Marshall, M 2009, 'Climate changes Europe's borders – and the world's', *New Scientist*, 27 March 2009, viewed 5 March 2019, <https://www.newscientist.com/article/dn16854-climate-changes-europes-borders-and-the-worlds/>.

Morris, ME 1990, *The icemen*, Grafton-Collins, London.

Parker, J & Lieber, S 2009, *Underground*, Image Comics, Berkeley.

Preisler, J 2001, *Cold war. Tom Clancy's 'Powerplay', no. 5*, Penguin, London.

Reilly, M 1998, *Ice station*, Pan Macmillan Australia, Sydney.

Reeves-Stevens, J & Reeves-Stevens, G 1998, *IceFire*, Pocket-Simon & Schuster, New York.

Sörlin, S 2015, 'Cryo-history: narratives of ice and the emerging Arctic humanities' in B Evengård, Ø Paasche & J Nymand Larsen (eds), *The new Arctic*, Springer, Cham, pp. 327–339.

Tabor, JM 2013, *Frozen solid*, Ballantine, New York.

Trexler, A 2015, *Anthropocene fictions: the novel in a time of climate change*, University of Virginia Press, Charlottesville and London.

Trexler, A 2012, 'Novel climes: Anthropocene histories, Hans-Jörg Rheinberger's *Trace*, and Clive Cussler's *Arctic Drift*', *Oxford Literary Review*, vol. 34, no. 2, pp. 295–314.

Turner, A 2016, *Out of the ice*, Simon & Schuster, Sydney.

White, G 2011, *Climate change and migration: security and borders in a warming world*, Oxford University Press, Oxford.

Woodhead, P 2015, *Beneath the ice*, Arrow-Random House, London.

7 Listening 'at the sea ice edge'

Compositions based on soundscape recordings made in Antarctica

Carolyn Philpott

Introduction

Since the earliest days of human encounters with the frozen continent, Antarctica has frequently been described by visitors as a relatively 'silent' place (Griffiths 2015, pp. 8–13; Philpott & Leane 2016, pp. 2–3). Despite this, hundreds of composers and sound artists have drawn creative inspiration from the far south for their musical and sonic art works, especially since the mid-twentieth century. The most prominent Antarctic-related compositions include Ralph Vaughan Williams's *Sinfonia Antartica* (1949–1952) and film score for *Scott of the Antarctic* (1948); Peter Maxwell Davies's *Antarctic Symphony* (2000), intended as a sequel to *Sinfonia Antartica*; and Nigel Westlake's *Antarctica* suite for guitar and orchestra (1992), which was derived from his original music for John Weiley's IMAX film *Antarctica* (1991). While Vaughan Williams and Westlake did not visit Antarctica to gain inspiration for their compositions, Maxwell Davies did (during the austral summer of 1997–1998). Another composer who has travelled to Antarctica is Paul Miller (also known as DJ Spooky that Subliminal Kid), who visited the continent in 2007–2008. In contrast to the more traditional orchestral works mentioned above, Miller produces experimental hip-hop music based on Antarctic climate-change data, which he regularly performs all over the world (Philpott [forthcoming]). In fact, over the past two decades, an increasing number of composers and sound artists have been travelling to Antarctica (mostly as part of national arts residency programmes) to experience its environment – and unique soundscapes – firsthand (Philpott 2016, p. 85).[1]

Most composers and sound artists who have visited Antarctica in recent years have employed sound recording technologies to document their journeys sonically and have subsequently created compositions based on their soundscape recordings. Typically, these compositions include biological sounds, such as vocalisations of penguins and seals (recorded both on the ice and underwater); and/or 'geophysical' (Pijanowski et al. 2011a, p. 1213) ambient sounds that emanate from the natural landscape, such as those created by wind, blizzards and ice cracking and calving. Since humans first began visiting the continent, however, its soundscape has regularly included the three dimensions of soundscapes typically found on other continents: various combinations of what musician and naturalist Bernie

Krause describes as '*biophony* (the sounds produced by non-human organisms in a given habitat), *geophony* (natural acoustic effects of wind ... water, and movement of the earth), and *anthrophony* (human-generated sounds some of which are controlled, like music ... and language, while the vast majority of them are chaotic and/or incoherent – sometimes referred to as noise)' (Krause 2016, p. 20).[2]

Increasingly, composers/sound artists have sought to capture anthropogenic (human) sounds from Antarctica within their compositions, often in combination with biological and/or geophysical sounds (Philpott 2016, p. 84). For example, Andrea Polli has incorporated anthropogenic sounds (as well as 'sonifications' of weather data)[3] from Antarctica in her *Sonic Antarctica* (2009) project. Craig Vear has also utilised both human and mechanical sounds (in addition to biological and geophysical sounds) in his CD/DVD/book project, *Antarctica: Musical Images from the Frozen Continent* (2012), based on recordings made in Antarctica in 2003–2004. Such efforts acknowledge the presence of human beings in the far south, as well as the reliance on human-made infrastructure and resources that is integral to the Antarctic experience for those who visit the continent for extended periods, including creative artists who undertake residencies there. By incorporating anthropogenic sounds as they exist in the natural soundscape (rather than editing them out), such artists aim to represent their sonic experiences in Antarctica, and their observations of human impacts on the region, as fully as possible.

Significantly, Antarctic-related musical and sonic art works reach vast audiences. This is especially true of Antarctic soundscape-based compositions, which are often heard in films and television programmes, in Antarctic exhibits at art galleries and museums, and through sound recordings (available in hard copy and/or online). Such works can provoke immediate and profound emotional responses from listeners. In this way, these works can enable diverse audiences to engage with Antarctica and to *feel* a sense of connection to the place – perhaps enhancing awareness of, and encouraging advocacy for, its environment in the process.

Some composers and sound artists who produce compositions in relation to Antarctica are acutely aware of the capacity for their works to inform audiences and to promote action about human-induced environmental change.[4] Similarly, scientists are increasingly recognising the role of the arts in enhancing public knowledge of important issues affecting the planet. As ecologist Mark Moffett has pointed out, 'Modern ecologists may have reached a limit on how effectively they can convey messages to the public, and they may now need to draw upon the emotional vibrancy offered by the arts' (quoted in Chadabe 2016, p. viii).

Musicologists and other scholars working in the field of ecomusicology have also started to grapple with questions relating to music and the Anthropocene. Leading researchers in the field Aaron Allen, Jeff Titon and Denise Von Glahn (2014, p. 7) argue that, 'The environmental crisis is not just a crisis of science ... but also a crisis of culture (failed thinking), so we need to muster all possible humanistic and scientific resources in order to imagine, understand and confront it'. Yet, to date, few studies within this research area have considered Antarctic-related topics. This is despite the southern continent's centrality to climate change issues – both as a source of information (through ice cores) and of potential sea-level rise (due to melting ice sheets) – and its role in inspiring some of the most significant composers of the last century, as well as an increasing number of contemporary composers and sound artists. It is this lacuna in the literature that this chapter – and much of my research more broadly – seeks to address.

Despite its attractiveness to composers and sound artists, Antarctica presents one of the most problematic environments on the planet for recording sound – just as it does for carrying out many human activities. The weather is highly changeable and often extreme, and animal life can be sparse and unpredictable: the continent thus offers unique challenges for even the most experienced sound artists to navigate – physically, technologically and creatively (Quin 1998). Two who are leading the way in this area are US-based composer Douglas Quin and Australian sound artist Philip Samartzis. Both artists have made significant contributions to this field, not only in terms of the quantity and quality of Antarctic-related works they have produced, but also through the innovative approaches they have developed to capture the sounds of this place and present them to broad audiences.

This chapter examines the contributions of Douglas Quin and Philip Samartzis to Antarctic-based sound art. The following sections introduce these artists individually and discuss specific examples of their Antarctic-related projects, focusing on compositions that incorporate sounds from underwater, as well as from above the ice (including, in Samartzis's case, anthropogenic sounds). The chapter builds upon and extends previous research I have published in *Organised Sound* and *The Polar Journal* (Philpott 2016; Philpott & Samartzis 2017), by examining in greater detail and breadth the Antarctic-based work of these two major sound artists, including numerous compositions that have not been discussed in the literature before. In doing so, my research sheds light on the role such works can play in enhancing public knowledge of a place that is beyond the reach of most humans – yet susceptible to anthropogenic change – while at the same time providing audiences with aesthetically pleasing listening experiences. In these ways, the chapter operates within the interdisciplinary space between the fields of ecomusicology and Antarctic studies, drawing together these otherwise relatively remote areas of intellectual inquiry.

Douglas Quin

Douglas Quin is an internationally acclaimed composer and sound designer who is also an Associate Professor in the Television, Radio and Film Department at the Newhouse School of Public Communications at Syracuse University, New York (Figure 7.1). In three decades of recording natural soundscapes, Quin's fieldwork has taken him all over the planet – in his own words, from 'Antarctic ice to Arctic tundra and from African savannah to Amazon rainforest' (Quin n.d.). His catalogue of original recordings of endangered animals and threatened habitats is one of the largest and most original yet created. Quin's recordings have been used by scientists, zoos and museums around the world for research and educational purposes, and he has also contributed to the sound design for significant feature films, including *Jurassic Park*

Figure 7.1 Douglas Quin recording Weddell seals in Antarctica
Photograph by James H. Barker, used with permission

III (2001), *Lord of the Rings: The Two Towers* (2002) and *Where'd You Go, Bernadette* (2019).

Quin has spent three separate summer seasons in Antarctica – in 1996, 1999 and 2000 – and was the first composer to travel there as part of the United States' National Science Foundation's Artists and Writers Program (Quin 2015, pers. comm., 1 April; National Science Foundation, n.d.). From these visits, he accumulated an extensive Antarctic sound library. His raw field recordings from the ice can be considered aural documentaries of his experiences there: they provide detailed sonic records of the places he visited at specific moments in time. Moreover, these recordings have provided a rich body of source material for an impressive catalogue of works.

Among the first creative outputs to stem from Quin's field trips to the southern continent was his ground-breaking soundscape compact disc, *Antarctica*, released by Miramar Recordings as part of Bernie Krause's 'Wild Sanctuary' series in 1998 (Quin 1998). This compact disc features soundscapes recorded by Quin during his first visit to Antarctica, including sounds of Emperor penguins, Weddell seals (both on land and underwater) and Adélie penguins, as well as sounds produced by the Canada Glacier and by wind harp installations Quin had set up in Taylor Valley. It concludes with a long piece recorded at the sea ice edge, which takes the listener deep below the surface, into an underwater community that is teeming with life.

Although Quin's *Antarctica* was not the first compact disc to incorporate sounds recorded in Antarctica,[5] it was among the first to focus entirely on presenting soundscapes from this region. Furthermore, the techniques Quin used to record sounds there – such as employing a multi-headed array of hydrophones (underwater microphones) to enable stereo/surround listening experiences – were arguably much more innovative and sophisticated than those used previously (Quin 1998).

His underwater recordings of Weddell seals, for example, provide richly detailed and unique aural snapshots of the sounds these enormous seals make beneath the ice. The sounds Quin captured are produced by male seals as they patrol their underwater territories, known as 'maritories' (Quin 2015). These seals have an extraordinary variety of vocalisations – researchers have classified twelve types of call with thirty-four discrete phrases – and their frequency range is impressive, reaching well beyond the levels of human hearing. Quin's 1996 recordings of these seals underwater remain striking, original and highly influential in the world of sound art. To make these recordings, he set up hydrophones at three sites, approximately fifty metres apart, and lowered the hydrophones to depths of twenty metres through holes drilled in sea ice two metres thick (Quin 2015, pers. comm. 22 January; Quin 1997). In an interview with Israeli composer and pianist, Jonathan Bar Giora, for the documentary film magazine, *Takriv* [Close-Up], Quin explained this process further:

> Recording the Weddell seals was a challenge and required a lot of plan-ning. ... I had help drilling through more than two metres of sea ice and camped out in my expedition tent for a few weeks making recordings. ... my first experience of hearing them was lying in my sleeping bag on the ice floor of my tent hearing and feeling the sound beneath me.
>
> (Quin 2015)

For the track 'Weddell Seals', Quin edited his raw sound recordings of the seals underwater only minimally. The piece features a remarkable array of sounds, perhaps the most notable of which are the seals' dramatic descending *glissandi*, which typically span more than an octave, overlap one another and are punc-tuated by a variety of chirps, clicks, thumps and other percussive effects. When Quin first heard these sounds, he could not believe his ears: they reminded him of vintage electronic music of composers such as Karlheinz Stockhausen (Quin 2015).[6] Later, he recognised parallels between the long, descending calls of the Weddell seals and the sounds produced by natural radio phenomena, known as 'whistlers', and noted similarities in the frequency ranges, shape, timbre and overlapping qualities of the two different types of sounds (Quin 1997, pp. 7–8). Although emanating from different sources, the two sound types are markedly similar, and the fact both are present in the south polar region led Quin to question whether Weddell seals can '"hear" or sense whistlers in ways that we do not yet understand' (p. 8).

Sounds produced by Weddell seals underwater can also be heard in the final track of *Antarctica*, 'At the Sea Ice Edge', in which they are showcased alongside underwater vocalisations of orcas and leopard seals. These, and the other tracks on this compact disc, provide excellent examples of context-based compositions that are both richly informative and aesthetically pleasing. Each piece has resulted from Quin's focused, in-place listening, his skill in capturing his sonic experiences, and his commitment to sharing the sounds he encountered in Antarctica with others.

In another of Quin's earliest Antarctic-related projects, the manner in which his sound recordings were presented was entirely novel – as well as highly ambitious. During his 1999 visit to the ice, also supported by the NSF's Artists and Writers Program, Quin created a live satellite sound installation and broadcast from the US-operated Palmer Station especially for the turn of the new millennium. Commissioned by the Studio Akustische Kunst [Studio Acoustic Art] at West German Radio (WDR) in Köln, this composition was titled *Die Jahrtausendwende: Live aus der Antarktis* [The Millennium: Live from the Antarctic]. It comprises pre-recorded sounds Quin had gathered over the preceding weeks mixed with live sounds from an array of two hydrophones (set up to record seals, whales and ice); a wind harp (to capture ambient wind); a Very Low Frequency (VLF) transducer (to record atmospheric whistlers and other phenomena from space); and a microphone capturing Quin's narration, including quotations from Ernest Shackleton's account of the ill-fated *Endurance* expedition of 1914–1917 (Quin 2000a and 2000b). These sounds were processed by Quin and transmitted live via satellite to WDR for broadcast on New Year's Day, 2000. Although Quin found the experience of creating the work live 'a bit hair raising with all the Y2K scare and just being at the very edge of the satellite network capability' (Quin 2014, pers. comm. 5 December), the event was a success, fulfilling the composer's intention for the project to 'celebrate the diversity of life and the wonder of the Antarctic' and bring 'a distant place into an immediate space through the medium of radio' (Quin 2000a).

Quin's sound recordings from his three visits to the ice have also been utilised in various films. For his third trip to Antarctica, he served as the location sound recordist for the Public Broadcasting Service (PBS) 2003 *Nature* documentary, *Under Antarctic Ice*, narrated by Hilary Swank. A few years later, he took on the roles of principal sound designer and sound mixer for Werner Herzog's Oscar-nominated documentary, *Encounters at the End of the World* (2008). Whereas *Under Antarctic Ice* required him to conduct further fieldwork in Antarctica, this was not necessary when it came to *Encounters at the End of the World*. Much of its footage was shot on location in Antarctica, and many of its sounds were drawn from Quin's existing archive of Antarctic recordings. As he explained to Jonathan Bar Giora:

> There was no need for me to go down [to Antarctica] again; we had most of what we needed for ambiences, and specific sound cues, planned in anticipation of shooting: everything from snowmobiles, to planes, helicopters and whatever environmental and creature audio we might need.
>
> (Quin 2015)

In the soundtrack for *Encounters at the End of the World*, Quin's 'real' sounds from Antarctica are juxtaposed with excerpts from Eastern Orthodox liturgical/church music and Bulgarian choral music. Given that most Western audiences would not be particularly familiar with these musical styles, their presence in the soundtrack helps, in Quin's words, to 'give the audience a visceral and emotional connection to the story' and 'imbue certain scenes with a sense of majesty and mystery' (Quin 2015). This is especially so in the underwater scenes, filmed under the ice. Describing one of these scenes, Herzog likens the divers to 'priests preparing for mass. Under the ice the divers find themselves in a separate reality where space and time acquire a strange new dimension. Those few who have experienced the world under the frozen sky often speak of it as, "going down into the cathedral"' (quoted in Quin 2015). In order to frame such otherworldly scenes in a realistic context, they are typically bracketed with Quin's 'real-life' sounds from his Antarctic field recordings. In this way, Quin's soundscape recordings help to situate Herzog's footage within a realistic sonic context, and 'bring the audience into the space literally ... without compromising the poetry and beauty' of the scenes (Quin 2015). Similarly, Quin's role as 'Antarctic Field Recordist' for the American comedy-drama film, *Where'd You Go, Bernadette* (2019) – based on the novel of the same name by Maria Semple and starring Cate Blanchett, Billy Crudup, Kristen Wiig and Judy Greer – did not require another journey to the ice. Instead, Quin again drew on his Antarctic sound archive to underpin scenes set there.

Sounds from Quin's Antarctic recordings have also been used away from their original contexts in films that are unrelated to the Antarctic. For example, sounds recorded during a 'hurricane-blizzard (Herbie) from within an empty million gallon fuel tank at McMurdo Station', which capture the creaking and groaning of the 'entire steel structure ... as it was pelted by ice in winds exceeding 160 km/hr', were not only used by Quin to great effect in *Encounters at the End of the World*, but also in the 'birdcage' scene in *Jurassic Park III*, in which they play a crucial role in generating nail-biting suspense. Additionally, he drew upon his Antarctic sound library when developing some of the dinosaur vocalisations within this film. To create realistic-sounding velociraptor vocalisations, for example, Quin and fellow sound designer Christopher Boyes used recordings of Emperor penguin calls, layered and edited with sounds from other animals. Through this process, they achieved a 'nasal quality' that matched scientific research suggesting that these dinosaurs might have produced sounds similar to birds (Quin 2018, pers. comm. 29 August).

Quin has also created various performative works incorporating his Antarctic sound recordings, including *Polar Suite*, composed for and performed by the Grammy Award-winning Kronos Quartet in 2011; the multimedia work *Aurora Passage*, composed in 2012 and based on the diaries of Bert Lincoln (an Able Seaman during the Australasian Antarctic Expedition, 1911–1914); and *Resonant Evidence* (2012–2015), a collaborative work with US-based composer Jay Needham derived from their collective polar field recordings (Philpott 2016, pp. 86–87; Quin & Needham 2012–2015). Another recent project, called *Paradise* (2016) – a multi-channel sound installation commissioned by the Society for New Music and developed in collaboration with Lorne Covington – provides an interactive experience in which visitors 'essentially compose a collage of virtual acoustic spaces drawn from the "natural" world', including Antarctica (Quin & Covington 2016). The installation's sound palette comprises almost 500 discrete audio clips, including underwater and glacial soundscapes, sounds of penguins, seals, whales and birds, wind and human voices, as well as myriad sounds from other regions. As listeners explore the installation, the resulting soundscape reflects their own presence and agency in art-making, as well as in the broader world. Through these and other works, Quin has enabled countless individuals, who will never visit the continent, to experience the sonic environment of Antarctica.[7] Even among those who are fortunate enough to go there, most are still unlikely to hear sounds from beneath the ice and surrounding waters of Antarctica in the clear, richly detailed and lengthy manner in which Quin has captured them.

In addition to the positive impacts his Antarctic work has had (and is continuing to have) on audiences around the globe, Quin is helping to inspire and advance scientific inquiries in relation to polar biology and glaciology. His work with the Weddell seals, for example, is contributing to scientific understanding of the species and its behaviour. In 2008, Quin co-authored with scientists an article in *Polar Biology* that reported the results of a study examining differences in underwater male Weddell seal trills at various locations around the perimeter of the Antarctic continent (Terhune et al. 2008). Quin contributed his Weddell seal recordings to this research, and also helped the scientists to design the scope of the study (Quin 2017, pers. comm. 24 August). Overall, his contributions were vital in shaping and informing this research, enabling the team to conclude that, 'It is likely that Weddell seals exhibit regional fidelity and never venture very far from their natal location' (Terhune et al. 2008, p. 679).

Quin's work has also inspired the research of American glaciologist Erin Pettit into the sounds of ice and what they can tell us about glaciers, as well as how they impact upon the behaviour of marine mammals.[8] While it is fairly common for science to influence art, or for art to provide a conduit for the communication of science to the public, it is not so usual to witness art or artists directly informing the work of scientists and contributing to scientific knowledge about the natural world. In this way, Quin's contributions are particularly significant. His work bridges the art–science divide in a way that benefits and extends both areas and which has the capacity to influence other researchers and arts practitioners working at the art–science nexus. Quin's work is already having an impact on other contemporary sound artists, one of whom is the Australian artist-academic, Philip Samartzis (Samartzis 2016a, p. 6).

Philip Samartzis

Like Douglas Quin, Philip Samartzis combines academic research and teaching with art practice. He is Coordinator of Sound in the School of Art at RMIT University in Melbourne, and his Antarctic-based work forms one aspect of a broader creative output that focuses on sound, art and the environment. Between 2010 and 2012, for example, Samartzis studied three Indigenous settlements in the

Kimberley region of Western Australia to sonically document the social and environmental conditions of remote communities. He is also the co-founder and artistic director of the Bogong Centre for Sound Culture (B-CSC), located in Alpine National Park in north-east Victoria, which supports a range of projects investigating the history and ecology of wilderness areas.

Samartzis has twice been awarded the Australian Antarctic Division Arts Fellowship (2009 and 2015), and has used these fellowship opportunities to document the effects of extreme climate and isolation on remote research stations in Eastern Antarctica and Macquarie Island, and on the research vessel *Aurora Australis* (Figure 7.2). Deeply inspired by the work of Douglas Quin and others who have 'interrogate[d] animal communication and behaviour, environmental conditions, and the transformative effects of climate and weather' (Philpott & Samartzis 2017, pp. 340–341), Samartzis has also sought to follow his own innovative path by focusing primarily on observing and documenting the 'effects of extreme climate and weather events upon various environments, communities and infrastructure' in Antarctica (quoted in Philpott 2016, p. 90). As such, many of his recordings from these trips incorporate anthropogenic sounds from Antarctic stations and the icebreaker *Aurora Australis*, as well as sounds from the natural environment.

Samartzis's Antarctic field recordings have formed the basis of numerous compositions which have been disseminated widely through performances at festivals,

Figure 7.2 Philip Samartzis in 'Iceberg Alley', Antarctica, March 2010
Photograph by Ian Aitkinson, used with permission

exhibitions, radio broadcasts and audio releases online and/or on compact disc. Sound recordings of six of his Antarctic compositions, for example, are provided on compact discs inside his book, *Antarctica: An Absent Presence* (2016). The compositions included with this book capture sounds recorded on board the research vessel *Aurora Australis* (in 'Icebreaker', 2015); sounds from the Australian base Davis Station (in 'Davis Station', 2015); recordings of a Medium Frequency Spaced Array (MFSA) radar used to measure upper atmospheric conditions (in 'At the End of Night', 2012); the sounds 'voiced by Antarctica's ice shelves, glaciers, icebergs and sea ice' (in 'Crush Grind', 2013) (Samartzis 2016a, p. 176); sounds Samartzis encountered on Macquarie Island (in 'Macquarie Island', 2013); and a piece derived from the sonification of digital data generated by auroral activity produced over one calendar year ('Aurora Australis', 2012) (Philpott & Samartzis 2017).

Two of these pieces, 'Crush Grind' and 'Macquarie Island', place the spotlight on 'geophysical' sounds that are prominent in the Antarctic and subantarctic (including those produced by different forms of ice and the effects of wind on natural and built environments, respectively).[9] Others focus more explicitly on the human presence in the far south by featuring sounds made by humans and/or infrastructure and equipment designed to support human life and activities on the continent. For instance, as its title suggests, the composition 'Icebreaker' presents some of the myriad sounds produced by the *Aurora Australis* as it navigates its way through the Southern Ocean and ice at the perimeter of the continent. These include the blasting of the *Aurora*'s horn at the opening of the piece, mechanical sounds from its engines, and the many different creaking, groaning, vibrating and crushing effects generated as it comes into contact with, and attempts to break through, various forms and thicknesses of ice. Similarly, although sections of 'Davis Station' expose the listener to the geophysical and biological sounds of the natural environment surrounding the most southerly of Australia's Antarctic bases (including water lapping at the coastline and vocalisations of Weddell seals on land), these serve only as a frame for the many anthropogenic sounds (from helicopters, trucks and generators, as well as human voices) that dominate the piece. Overall, this composition provides a stark reminder that within this remote wilderness, human beings have established infrastructure and resources to enable communities of scientists and support personnel to live and work, and that human-made sounds are now also a prominent feature of the soundscape in various parts of the continent.

Given that human presence in Antarctica is so closely connected with – and reliant upon – the scientific bases, it is not surprising that Samartzis found this aspect of the local soundscape (and his own experience in Antarctica) impossible to ignore. Indeed, he felt compelled to share these sounds through his compositions in order to represent the environment accurately. As he has acknowledged:

Life on the station is a compendium of natural and human sounds. Rather than trying to privilege one over the other I accept all sounds as being of equal interest, and true of the Antarctic experience ... [Through my sound art] I want the environment to reveal its own biases and dynamics with as little intervention as possible. In this way listeners are free to discover characteristics and behaviours interesting to them, presented within a framework of a naturally occurring set of interactions and exchanges. ... My role is to direct attention to places and actions that are often overlooked or disregarded through the aperture of sound recording ... Composition for me therefore is a product of selecting and sequencing sound recordings that accurately reveal the ecology of place.

(quoted in Philpott & Samartzis 2017, p. 347)

Through his efforts to map Davis Station and its surrounding environment in sound, Samartzis has been able to fulfil his aim of providing 'unique insight into the way anthropogenic, natural and environmental forces dynamically converge to shape the Antarctic experience' (2016a, p. 20). As Douglas Quin writes in his Foreword to *Antarctica: An Absent Presence*:

Philip's exquisite digital recordings are meticulously realised, crisply detailed and reveal true artistry from field to finished composition. Much like a painter will use different brushes or a photographer a variety of lenses, Philip employs a range of microphones and approaches in his recording. ... These instruments provide us with visceral and concrete sonic connections to place, materials, matter and objects. ... *Antarctica: An Absent Presence* is a remarkable journey in sound – one that is as lyrical as it is rich in descriptive detail and nuance.

(Quin 2016, pp. 8–10)

The book, *Antarctica: An Absent Presence*, provides further insights into Samartzis's Antarctic experiences that serve to contextualise the compositions. The text includes excerpts from the diary Samartzis kept during his first visit to the ice and is interspersed with many of the photographs he took to document his journey visually. According to the artist, 'the journal acts as a repository for displaced memories – real and imagined – chronicling the people, places and conditions I encountered, supplemented by descriptions of technical and artistic process. The format is designed to acknowledge the revered, almost religious place of the journal in Antarctic exploration' (quoted in Philpott 2016, p. 90). The images presented in the book again predominately reflect Samartzis's focus on human life on the continent and the effects of weather on the built station environment; they provide relevant visual imagery to support the text and soundscape compositions. As an extension of this project, in 2014, he composed an hour-long radio piece in French for France Culture (Samartzis 2014a), and an English-language version for ABC Radio National (Samartzis 2014b), also titled *Antarctica: An Absent Presence*. This composition comprises readings from Samartzis's diary set against a backdrop of relevant sounds from his journey, thereby providing an engaging audio journal of his visit to Antarctica. As a whole, the *Antarctica: An Absent Presence* project provides, in the composer's words, 'a compendium of observations using sound recordings and text formatted as a journal to convey an experience of Antarctica different to those presented through the lens of natural history, climate science or human exploration. It is a reimagining of a place marked by a strong sense of absence' (quoted in Philpott 2016, p. 90). While the place he describes might be 'marked by a strong sense of absence', as Samartzis's compositions reveal, the frozen south is far from silent.

The most recent outcome from Samartzis's engagement with Antarctica – and the first composition to emerge from his second visit to the continent in early 2016 – is a twenty-one-minute composition titled 'The Blizzard' (2017). Taking inspiration from Frank Hurley's 1912 photographs, *The Blizzard* and *Leaning on the Wind*, Samartzis aimed during this visit to document the effects of katabatic wind on Casey Station, another of Australia's Antarctic bases. This type of wind is particularly common at Casey Station due to its location at the base of Law Dome, a large ice dome that rises to a height of 1395 metres. This, according to the composer, frequently results in an 'erratic weather system ... making Casey the ideal location' for recording this phenomenon (Samartzis 2016b). He elaborates:

> *The Blizzard* is a multi-channel composition comprising sound recordings made on February 15, 2016, at Casey Station, Eastern Antarctica, of katabatic wind gusting up to 185 KPH. Katabatic wind is a low gravity wind that gains force as it travels down elevated slopes. When the cooler temperature of a katabatic mixes with the warmer temperature of the onshore wind, a very unstable weather system emerges. Katabatic wind is notable for the way it shapes the behaviour of sound within the built and natural environment. Its trajectory can push sound away or draw it closer. Its intensity can mask sound and its absence can heighten it. At its most ferocious it simply obliterates everything in its path. A collision with the built environment of Casey Station transforms katabatic wind into an intense series of ascending and descending pitches – a supercharged Aeolian harp. Inside the braced steel framed and insulated panel buildings pervades a silence that imposes a profound sense of isolation from the immediate environment. Outside the volatility is expressed through a variety of resonances emitted by miscellaneous surfaces and materials undergoing tremendous stress.
>
> (Samartzis 2017, pers. comm. 14 July)[10]

Samartzis has created a twenty-four-channel version of 'The Blizzard' for playing through a forty-loudspeaker system (first installed at the 2017–2018 Melbourne exhibition, *Super Field*). This setup allows listeners to experience a surround-sound version of the piece that places them sonically in the midst of the katabatic wind Samartzis had experienced at Casey Station.

Other creative outcomes from Samartzis's visits to Antarctica have appeared in exhibitions, symposia and/or performances in Argentina, Scotland, and the United States, as well as Australia. For example, his work appeared in the exhibition *Polar South: Art in Antarctica*, held at the Muntref Museum, National University of Tres de Febrero, Buenos Aires (2011); the Eleventh International Symposium on Antarctic Earth Sciences, Edinburgh (2011); the National Archives of Australia's exhibition,

Traversing Antarctica: The Australian Experience (2011–2014); and in a concert held as part of the *Balance–Unbalance* international conference in Arizona (2015). Furthermore, he has curated Antarctic-related festivals and exhibitions, such as the *Antarctic Convergence Festival*, which toured through Australia and New Zealand in 2012, showcasing the works of artists and composers who have travelled to Antarctica; and *Polar Patterns*, a group exhibition exploring the eco-acoustic characteristics and atmospheric effects of Antarctica, held at RMIT University, Melbourne, in early 2017. Through these and other activities, Samartzis's work has reached broad audiences and made a 'unique and important contribution to how Antarctica is understood and appreciated through the aperture of sound art' (Philpott & Samartzis 2017, p. 349).

Conclusions

The Antarctic-related works of composers and sound artists such as Douglas Quin and Philip Samartzis are important for several reasons. Perhaps most obviously, these works make a significant contribution to the fields of sound art and music by employing innovative approaches to capturing places in sound that can be used in a range of creative contexts, from performances in concert halls and at festivals, to exhibits in art galleries, museums and zoos, through to film soundtracks. Beyond their contributions to music and sound art, however, these works impact significantly upon the general public by allowing people from diverse backgrounds and geographical locations to hear the unique sounds of the far south, and learn more about the region, without actually having to *go there*. As photographer Gary Kolb and sound artist Jay Needham have observed, such work can influence the 'public conception of "unknowable" spaces that are beyond the reach and view of the average person' (2014, p. 7). Similarly, scholars in the field of soundscape ecology argue that listening to soundscapes can also inspire 'appreciation, management and conservation of the organisms and resources that create them' (Pijanowski et al. 2011a, p. 1227). This is particularly important in relation to Antarctica, given its prominent role in global climate processes, its susceptibility to environmental change, and the fact that the continent is beyond the 'reach and view' (and *earshot*) of most people.

Even for those members of the public who do visit Antarctica on tourist cruises, most will never experience sounds from underwater or from the scientific stations and vessels as captured in the recordings of Quin and Samartzis. In fact, at least one recording by Quin captures the sounds of a place in Antarctica – an ice cave – that no longer exists (Philpott 2016, p. 87). As time goes on, it might well be the case that listening to soundscape recordings is our only means of hearing certain species and places that have been affected irrevocably by environmental change. As Pijanowski and others note, 'sound is a fundamental property of nature and because it can be drastically affected by a variety of human activities [it is likely that] … recordings made today will become tomorrow's "acoustic fossils", possibly preserving the only evidence we have of ecosystems that may vanish in the future because of a lack of desire or ability to protect them' (Pijanowski et al. 2011b, pp. 203, 213). Antarctic soundscape recordings, therefore, hold significance as sonic records or artefacts, in addition to providing the basis for aesthetically pleasing listening experiences that allow us to engage – in deep and powerfully affecting ways – with the depths and surfaces of Antarctica.

By examining the contributions of Douglas Quin and Philip Samartzis to Antarctic-based sound art, this chapter has aimed to increase awareness and enhance understanding of their work. It is hoped that readers will be inspired to seek out recordings of the compositions discussed here – as well as other Antarctic-related sonic and musical works – and *listen* to them with an open mind. The research presented here augments previous studies I have undertaken on Antarctic-based soundscape compositions. It extends that work by focusing on the Antarctic-related outputs of these two major figures – and specifically on their pieces that incorporate sounds from underwater, as well as from above the ice, including from anthropogenic sources. As the first scholarly essay to draw attention to these particular works, the chapter advances knowledge in – and forges connections between – the fields of ecomusicology and Antarctic studies. Ultimately, it also prompts readers to consider what Antarctica's sounds can tell us about the so-called 'silent continent' and how hearing them can challenge our assumptions about this remote region and potentially make us feel more connected to it – whether we are there in person or listening from a distance.

Notes

1 Others include Lawrence English, Cheryl Leonard, Jay Needham and Ian Tamblyn.
2 See also Krause 2013.
3 Sonification involves mapping or converting data to sound for the purpose of conveying information about the data to listeners.
4 See, for example, Polli 2016; Philpott 2018, pp. 39–50.
5 See, for example, ABC Music 1993 and Beasts of Paradise 1995.
6 German composer Karlheinz Stockhausen (1928–2007) pioneered the use of electronic music, among other innovations, in the mid-twentieth century. See Toop 2001.
7 In 2010, for example, Quin released an LP titled *Fathom* (on the label Taiga), which features two extended underwater soundscapes from Antarctica (as well as two from the Arctic).
8 Pettit acknowledges Quin's influence in her sole-authored article, Pettit 2012, p. 121. See also Pettit et al. 2012, pp. 104–105; and Pettit et al. 2015, pp. 2309–2316.
9 As Samartzis writes in his description of 'Crush Grind', the sounds produced by Antarctica's ice as captured in this composition 'contest one of the great misconceptions about the continent … that it is a place delimited by a rigid and mute set of conditions. Yet concealed within the frozen veil of ice is a startling aggregation of sound which demonstrates how remarkably protean the continent actually is' (2016a, p. 176).
10 See also Samartzis 2016c. Another Australia-based sound artist, Lawrence English, has also recorded the sounds of blizzards in Antarctica. See Philpott 2016, p. 89.

References

ABC Music 1993, *Antarctic Symphony*, CD, ABC Music 514639–514632.
Allen, AS, Titon, JT & Von Glahn, D 2014, 'Sustainability and sound: ecomusicology inside and outside the academy', *Music & Politics*, vol. 8, no. 2, pp. 1–26.

Beasts of Paradise 1995, *Gathered on the Edge*, CD, City of Tribes COTCD 008.

Chadabe, J 2016, 'Foreword' in F Bianchi & VJ Manzo (eds), *Environmental sound artists: in their own words*, Oxford University Press, Oxford, pp. vii–viii.

Griffiths, T 2015, 'Introduction: listening to Antarctica' in B Hince, R Summerson & A Wiesel (eds), *Antarctica: music, sounds and cultural connections*, ANU Press, Canberra, pp. 8–13.

Herzog, W (dir.) 2008, *Encounters at the end of the world*, motion picture, Discovery Films.

Jackson, P (dir.) 2002, *Lord of the rings: the two towers*, motion picture, New Line Cinema.

Johnston, J (dir.) 2001, *Jurassic Park III*, motion picture, Amblin Entertainment/Universal Pictures.

Kolb, G & Needham, J 2014, 'Antarctic dreams', *Exposure*, vol. 47, no. 2, pp. 4–7.

Krause, B 2016, 'Biophonic sound sculptures in public spaces' in F Bianchi & VJ Manzo (eds), *Environmental sound artists: in their own words*, Oxford University Press, Oxford, pp. 19–25.

Krause, B 2013, *The great animal orchestra: finding the origins of music in the world's wild places*, Back Bay Books, New York.

Linklater, R (dir.) 2019, *Where'd you go, Bernadette*, motion picture, Annapurna Pictures.

National Science Foundation, *Antarctic artists & writers program — Past participants*, viewed 3 March 2019, <https://www.nsf.gov/geo/opp/aawr.jsp>.

Pettit, E 2012, 'Passive underwater acoustic evolution of a calving event', *Annals of Glaciology*, vol. 53, no. 60, pp. 113–122.

Pettit, E, Lee, K, Brann, J, Nystuen, J, Wilson, P & O'Neel, S 2015, 'Unusually loud ambient noise in tidewater glacier fjords: a signal of ice melt', *Geophysical Research Letters*, vol. 4, pp. 2309–2316.

Pettit, E, Nystuen, J & O'Neel, S 2012, 'Listening to glaciers: passive hydroacoustics near marine-terminating glaciers', *Oceanography*, vol. 25, no. 3, pp. 104–105.

Philpott, C [forthcoming], 'Mixing ice: DJ Spooky's musical portraits of the Arctic and Antarctic' in C Philpott, M Delbridge & E Leane (eds), *Performing ice*, Palgrave Macmillan, London.

Philpott, C 2016, 'Sonic explorations of the southernmost continent: four composers' responses to Antarctica and climate change in the twenty-first century', *Organised Sound*, vol. 21, no. 1, pp. 83–93.

Philpott, C 2018, 'Promoting environmental awareness through context-based composition', *Organised Sound*, vol. 23, no. 1, pp. 39–50.

Philpott, C & Leane, E 2016, 'Making music on the march: sledging songs of the "Heroic Age" of Antarctic exploration', *Polar Record*, vol. 52, no. 6, pp. 698–716.

Philpott, C & Samartzis, P 2017, '*At the End of Night*: explorations of Antarctica and space in the sound art of Philip Samartzis', *The Polar Journal*, vol. 7, no. 2, pp. 336–350.

Pijanowski, BC, Farina, A, Gage, SH, Dumyahn, SL & Krause, BL 2011a, 'What is soundscape ecology? An introduction and overview of an emerging new science', *Landscape Ecology*, vol. 26, no. 9, pp. 1213–1232.

Pijanowski, BC, Villanueva-Rivera, LJ, Dumyahn, SL, Farina, A, Krause, BL, Napoletano, BM, Gage, SH & Pieretti, N 2011b, 'Soundscape ecology: the science of sound in the landscape', *BioScience*, vol. 61, no. 3, pp. 203–216.

Polli, A 2016, 'Sonifications of global environmental data' in F Bianchi & VJ Manzo (eds.), *Environmental sound artists: in their own words*, Oxford University Press, Oxford, pp. 3–7.

Public Broadcasting Service 2003, 'Under Antarctic Ice', DVD.

Quin, D n.d., 'Doug Quin', *EarthEar*, viewed 8 August 2017, <http://earthear.com/quin.html>.

Quin, D 1997, 'Antarctica: austral soundscapes', viewed 10 August 2017, <http://www.dqmedia.com/articles/austral.pdf>.

Quin, D 1998, *Antarctica*, CD, Miramar 09006-23113-2.

Quin, D (Audionomad) 2000a, 'Die Jahrtausendwende: Live aus der Antarktis', *Sound-Cloud*, viewed 11 August 2017, <https://soundcloud.com/audionomad/die-jahrta usendwende-live-aus>.

QuinD, 2000b, 'Live soundscape broadcast from Palmer Station, Antarctica (December 31, 1999 – January 1, 2000)', unpublished diagram of installation and on-site setup.

Quin, D 2010, *Fathom*, LP, Taiga Records.

Quin, D 2015, *Douglas Quin interviewed by Jonathan Bar Giora for Takriv* [Close-Up].

Quin, D 2016, 'Foreword', in P Samartzis, *Antarctica: an absent presence*, Thames & Hudson, Melbourne, pp. 8–10.

Quin, D & Covington, L 2016, 'Paradise, an interactive sound installation', brochure.

Quin, D & Needham, J 2012–2015, *Resonant evidence*, *SoundCloud*, viewed 17 August 2017, <https://soundcloud.com/jay-needham-1/resonant-evidence-by-jay-needham -and-douglas-quin#t=0:00>.

Samartzis, P 2014a, 'ACR – Antarctique, une présence absente', podcast, *L'atelier de la création*, France Culture, viewed 17 August 2017, <http://www.franceculture.fr/em ission-l-atelier-de-la-creation-acr-%E2%80%93-antarctique-une-presence-a bsente-2014-12-04>.

Samartzis, P 2014b, 'Antarctica, an absent presence', podcast, *Soundproof*, ABC Radio National, viewed 17 August 2017, <http://www.abc.net.au/radionational/programs/soundproof/antarctica-an-absent-presence/5930156>.

Samartzis, P 2016a, *Antarctica: an absent presence*, Thames & Hudson, Melbourne.

Samartzis, P 2016b, 'Katabatic winds inspire sound rendition of Antarctic experience', *Australian Antarctic Magazine*, vol. 30, viewed 18 August 2017, <http://www.anta rctica.gov.au/magazine/2016-2020/issue-30-june-2016/antarctic-art/katabatic-wind s-inspire-sound-rendition-of-antarctic-experience>.

Samartzis, P 2016c, 'Sounds from the home of The Blizzard', *Real Time*, no. 131, viewed 18 August 2017, <http://www.realtimearts.net/article/issue131/12195>.

Terhune, JM, Quin, D, Dell'Apa, A, Mirhaj, M, Plötz, J, Kindermann, L & Borne-mann, H 2008, 'Geographic variations in underwater male Weddell seal trills suggest breeding area fidelity', *Polar Biology*, vol. 31, no. 6, pp. 671–680.

Toop, R 2001, 'Stockhausen, Karlheinz', *Grove Music Online, Oxford Music Online*, Oxford University Press, doi:10.1093/gmo/9781561592630.article.26808.

Vear, C 2012, *Antarctica: musical images from the frozen continent*, CD/DVD/book, Gruenrekorder, Frankfurt.

Weiley, J (dir.) 1991, *Antarctica*, motion picture, AFFC/Heliograph Productions/ Museum of Science and Industry Chicago.

8 Save the penguins
Antarctic advertising and the PR of protection

Hanne Nielsen

In early 2002, the Larsen B Ice Shelf on the eastern side of the Antarctic Peninsula collapsed with dramatic speed. The break up was photographed via NASA's MODIS satellite, resulting in a series of images that have since been combined into a time-lapse series. Widely disseminated, these images provide dramatic visuals for the narratives of climate change that continue to circulate today (Plester 2016). Never before had such a large area disintegrated so rapidly (Lindsey 2002), and never before had visual imagery of this sort of event been imbued with such strong cultural meaning. Cultural critic Judith Williamson (2010, n.p.) argues that 'imagery circulating in our societies has a profound effect on our ability to grasp the world as it is, and imagine how it might be'. In the case of the Larsen B Ice Shelf, the imagery of the collapse (Figure 8.1) was transmitted across the globe, used to illustrate news items and reproduced in print and digital forms. It came to symbolise a fragile climate system that was already melting, cracking and disintegrating on a very human timescale.

Ice is understood to be particularly vulnerable to anthropogenic climate change, largely because melting can be seen (when observed over a period of time), while many other effects of climate change remain invisible (Carvalho 2010, p. 489). As a result, ice often serves as a visual metaphor for change. The visibility of the Larsen B collapse is just as significant as the short timeframe over which it occurred. Kathryn Yusoff (2005, p. 387) has written that 'one can experience the compression of the reality of time and space' within this series of images, where 'the predominant narration becomes about witnessing the spectacle of change'. Indeed, the time-lapse series makes visible 'the moment when disaster strikes' (p. 387). This visibility means that the event has had a long afterlife in popular culture. Imagery of the cracking Larsen B Ice Shelf has been recycled in the opening of the US disaster film *The Day After Tomorrow* (Emmerich 2004), inspired artworks,[1] and appeared on t-shirts: online clothing company Zazzle (n.d.) promotes the *Larsen B Ice Shelf Collapse (Picture Earth) Shirt* as a gift 'that will surely get others talking about the consequences of global warming. [It is the] Perfect educational science gift for all fans and advocates of the existence of Antarctica, the ice continent!' Such cultural products illustrate how melting ice has come to stand for a fragile climate system, with Antarctica providing a shorthand for vulnerable ecosystems everywhere. Comprised of ice and glaciers that are 'susceptible to cultural framing as

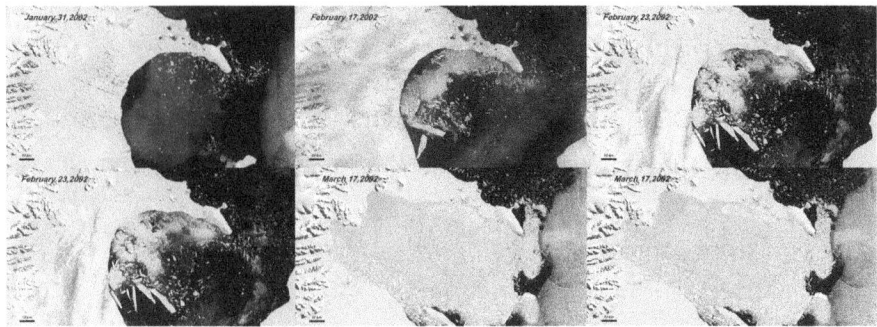

Figure 8.1 Collapse of the Larsen B Ice Shelf, 2002
Source: NASA/Goddard Space Flight Center, 'Collapse of the Larsen B Ice Shelf'

both dangerous and endangered landscapes' (Nüsser & Baghel 2014, p. 138), Antarctica is seen simultaneously as fragile and treacherous. The continent is cast as a place that both threatens humankind and needs to be protected from the effects of anthropogenic climate change. This dissonance is key to representations that call upon Antarctica to embody environmental ideas.

When it comes to the idea of fragility, the use of the Larsen B imagery to articulate the phenomenon of global change is just the tip of the metaphorical iceberg.[2] This chapter examines how the fragile connotations of Antarctic imagery have been put to use in advertising material. First, it addresses the various ways in which Antarctica has been popularly framed since humans first began to physically interact with the place. A discussion of the commercialisation of polar imagery is followed by an analysis of three international advertising campaigns: ABB's 'amazing what you save' (2002, 2005, 2008); Westpac's 'Equator Principles' (2003, 2008); and Diesel's 'Global Warming Ready' (2007) campaigns. In these case studies, advertisements for three very different industries are used as a proxy for accessing dominant attitudes towards the far south and to illustrate a contemporary face of the commercial history of Antarctica.

Mobilising multiple framings of Antarctica

Antarctica has been framed in multiple ways over the course of its human history. For the early sealers and whalers who headed south for the hunt during the nineteenth and twentieth centuries, the region was a place for profit. Interest in resources such as minerals, marine living resources – and even ice – has continued into the twenty-first century: the United Arab Emirates' 2017 announcement of plans to tow an iceberg north for drinking water is a recent iteration of this commercial interest (Express Web Desk 2017). The negotiations that took place throughout the 1980s relating to the unratified Convention for the Regulation of Antarctic Mineral Resource Activity (CRAMRA) demonstrate high-level awareness of Antarctica's resource potential. Rational use continues to be a central tenet of the Convention on the Conservation of Antarctic Marine Living Resources

(CCAMLR), which governs fishing activity in the Southern Ocean. The framing of Antarctica as a realm of commerce therefore has a long and varied history.

In recent decades, however, the rhetoric of profit has increasingly given way to that of protection. Article 2 of the Protocol on Environmental Protection to the Antarctic Treaty (Madrid Protocol) signed in 1991 designates Antarctica as 'a natural reserve, devoted to peace and science' (Secretariat of the Antarctic Treaty 1991). It protects the ecosystems of the far south, as well as the 'wilderness and aesthetic values' of the region; it also prohibits mining activity. Although the coupling of Antarctica with protection is comparatively recent, this is the dominant lens through which the continent is discussed in the Anthropocene.

At the same time, Antarctica has a long and enduring association with the twin themes of heroism and extremity. During the period of early land-based exploration (the 'Heroic Era' of 1895–1922), tales of exploration and images of figures battling blizzards to achieve geographic firsts cemented the popular connection between Antarctica and heroism. The continent has been cast as the ultimate testing ground for technologies such as vehicles and clothing, while the clichéd description of Antarctica as the coldest, driest, highest continent on Earth reinforces the idea of extremity. In more recent years, Antarctica has also been used metaphorically to stand in for any kind of challenge – the Shackleton Foundation's (2014) campaign headline, 'Everyone has an Antarctic. What's yours?', is but one example. Here, disadvantaged young people are encouraged to picture adversity as an icescape, and to triumph over the imagined polar environment, building upon established ideas of heroism, extremity and endurance.

The themes of purity and fragility act as a counterpoint to these tropes of extremity – rather than depicting Antarctica as a place for humans and machines to tackle the dangers of a hostile landscape, they transfer the vulnerability to the landscape itself. The far south has stronger associations with purity and fragility than other cryoscapes for a range of reasons.[3] Antarctica's exceptional circumstances – its unowned status, the absence of an indigenous population, its remoteness and short human history – contribute to its framing as particularly vulnerable. These factors mean that Antarctica is more easily understood as 'pure wilderness', more susceptible to human pollution/destruction. In recent years, this corrupting threat has been from anthropogenic climate change. In March 2015, the monthly global average concentration of CO_2 in the atmosphere surpassed four hundred parts per million, bringing the concentrations into unprecedented territory (NOAA 2015). The greenhouse effect is becoming ever stronger, leading to increasing temperatures in the atmosphere and the ocean. Climatologist Gavin Schmidt, from NASA's Goddard Institute for Space Studies, warns that, 'We are a society that has inadvertently chosen the double-black diamond run without having learned to ski first. It will be a bumpy ride' (NASA 2013).

Physically, politically and symbolically, then, the Antarctic region has global significance. Climate change has an impact upon ice, ocean and beyond, as the cryosphere acts as 'a fundamental control on the physical, biological and social environment over a large part of the Earth's surface' (Vaughan et al. 2013, p. 319). Changes in the far south contribute to feedback loops with implications for places much further afield; sea-level rise and changes in ocean circulation are but two areas 'where the processes in Antarctica are fundamentally important globally' (Rodger 2013, p. 324). This also has an impact upon how Antarctica is viewed, as the continent transitions from somewhere remote and invisible to part of the global climate system. Indeed, the melting ice makes the effects of climate change tangible. It turns the global and the political into the local and the personal, as Antarctic change is experienced as rising sea levels along faraway coastlines. At the same time, climate change is being 'reimagined as an ethical, societal, and cultural problem that poses new questions and reconfigures the geographic imaginaries of the world' (Yusoff & Gabrys 2011, p. 517). Within this reimagining, Antarctica has come to play a dominant role as a symbol for climate change, for fragility and for the threat of melting ice. This symbol is particularly prominent in media texts and advertisements.

Frozen imagery and 'ice-wash'

For most people, access to Antarctica is primarily through the mediation of news images, documentaries, social media and advertising. When the dominant imagery consists of calving icebergs and melting ice, this helps to frame Antarctica as a fragile place, vulnerable to change. In this way, it shares similarities with the Arctic, where icy imagery has been used discursively to stand for the same environmental issues, although the relationship between forms of media and the politics of Arctic climate change are far better documented (Christensen, Nilsson & Wormbs 2013). In the South Polar context, penguins and icebergs are often used to suggest that a company has strong environmental credentials. Such promises are not always fulfilled. In *Green Wash*, environmental commentator Guy Pearse highlights the 'gulf between the green revolution being advertised and the progress actually occurring' (2012, p. 246). 'Ice-washing' is the polar equivalent to Pearse's green wash, with the environmental message conveyed or suggested via images of glaciers and the cryosphere at large (Williamson 2010).

Several issues are associated with employing polar imagery for commercial purposes. First, the aim of most advertising is to promote consumption. Consumerism (and increased consumption), however, is one of the major factors behind environmental problems such as climate change (Mayell 2004). Advertisements that exhort consumers to buy a particular product in order to save the environment are rarely self-reflexive about their own role in environmental damage. Environmentally conscious consumers are not immune to these contradictions. Paul Simpson-Housley has written about how 'sometimes the cognitive and affective components (of a perception) clash' (1992, p. xv) – one might be seduced by polar imagery, despite knowing that its visual promise of protection will not be fulfilled, just as one can know that fire pollutes but still enjoy its warm glow. Second, there are particular issues inherent in using snow and ice to represent climate change, particularly change that manifests as warming. As Williamson has noted:

We see pictures of coldness, not heat, of glaciers, not droughts. Whatever the logical reason for these images, their constant presence, the cultural ubiquity of these frozen landscapes, functions as an imaginative denial of the real nature of climate change, and the situation faced by large parts of humanity today.

(2010, n.p.)

In this case, the symbolic power of snow and ice in the Antarctic renders invisible contemporary environmental issues in other parts of the globe. Such examples show that employing icy imagery in order to drive home an environmental message is not straightforward. As the cryosphere has entered into more and more conversations, it has attracted a wider range of symbolic meanings. As Nüsser and Baghel (2014, p. 150) put it, 'glaciers do not *just* melt; they are imbued with cultural, scientific, political and aesthetic meanings'. Ice is not just ice – rather, it carries with it a range of cultural connotations, specific to each time and place in which the imagery appears, leading Sverker Sörlin (2015, p. 327) to argue that 'we have reached a "cryo-historical" moment'.

The following section presents three advertisements as case studies in how Antarctica has been used as a symbol for both the polar environment and the more-than-human world at large. These examples are representative of a much larger body of advertisements that call upon the theme of Antarctic fragility for marketing purposes.[4] The three advertisements discussed here clearly illustrate how the rhetoric of protection has been deployed in a commercial context. They also signal that, far from being just a marginal mass of ice that is regularly left off the map, Antarctica plays an important role in the cultural imaginary of climate change. In all three cases, large companies capitalise on the symbolic value of the southern continent, transforming fragility into financial gain.

Saving ice: Antarctica and the rhetoric of protection

In 2002, robotics and automated technology company ABB's global campaign featured several dozen Adélie penguins on a large iceberg, accompanied by the question: 'Can you stop 50 million tons of CO_2 from happening?' (Hicks 2005).[5] The campaign was repeated in 2005, this time with the Adélies replaced by a close-up row of King penguins on the ice, and updated numbers – 68 million tons, up from 50 million tons (ABB 2005; West, Ford & Ibrahim 2012, p. 450). In 2008, the same King penguins made a reappearance in the MIT European Career Fair booklet, with the number of tons of CO_2 in question boosted to 100 million (ABB 2008). The reprise of the campaign suggests a positive reception; zooming in on the penguins in the second and third iterations suggests that these birds are effective at creating meaning in the advertisement. These penguins are not just penguins – rather, they embody an environmental message that has relevance from the poles to the job fairs of Europe. In this trade context, environmental priorities are presented as a selling point, and act as a counterpoint against the technological imagery or people in other advertisements throughout the booklet.

Penguins are often used to stand in for Antarctica. They are a recognisably southern-hemisphere species, associated – thanks to countless photographs, cartoons, emojis and other appearances in popular culture – with snow and ice. Even though the ABB advertisement features King penguins, which breed on the sub-Antarctic islands rather than the icy continent itself, they serve as shorthand for Antarctica.[6] The ABB penguins act on several levels of their 'complex and contradictory symbolic repertoire' (Leane & Pfennigwerth 2011, p. 40): they are used to represent the megafauna of Antarctica; to stand for the environment they live in and around; and to gesture towards the wider issue of climate change. The advertisement assumes that readers are aware of the process by which CO_2 emissions result in the melting of ice, thus endangering wildlife.

ABB's penguin advertisements have been described as part of an 'environmentally responsible campaign' (West, Ford & Ibrahim 2012, p. 450). They appeared at a time when big business was beginning to see the value in projecting an eco-friendly image. The 'amazing what you save' campaign was designed to highlight the company's new automated factory technology – 'a variable-speed drive unit'– that allowed factories to slow the emissions put out during quieter production periods, thus 'saving energy when the plant was not needed at full capacity production' (p. 450). The monetary savings were a prime selling point of the technology, with the variable-speed drives touted as 'major energy savers, environmentally friendly and a wise investment' (Savolainen 2004, p. 34). Environmental issues also translate back into profit when public opinion and the 'reflected green glow' (Pearse 2012, p. 18) of an energy-efficient line are factored in. If a particular product is seen as environmentally friendly, that has implications for the brand as a whole, because 'all commercials for a product also advertise the brand' (Pearce 2012, p. 37). The appearance of a polar landscape in this advertisement therefore has wider implications for the company than simply serving to market a single product: it creates an association between the ABB brand and environmental practices and imagery that endures far longer than the ephemeral advertisement itself.

The play on the words 'what you save' is pivotal to this ABB advertisement. Taken literally, the term 'save' is used in a monetary sense – cutting many millions of tons of CO_2 will reduce energy and business costs. However, thanks to the juxtaposition of text and image, the term 'save' can also be understood as referring to the penguins themselves, inviting environmental as well as economic connotations (Savolainen 2004). Penguins have been put to use in advertising imagery in a range of ways over the years, but are commonly used to 'act as synecdoches for pristine nature' (Leane & Pfennigwerth 2011, p. 30). When the penguins are used to stand in for Antarctica as a whole, with Antarctica in turn standing in for the world's environmental systems, the question of what is being saved gains even more significance. As a solution to this problem, consumers are exhorted to buy ABB products, thereby reducing their carbon footprints and saving money, while helping protect the environment. Antarctica functions as a backdrop to the ABB advertisement, but that backdrop brings salient environmental issues to mind. The advertisement therefore presents an earnest appeal to consumers that centres round the rhetoric of saving energy, saving money, saving penguins and saving the polar environment.

How to change to a globe: Westpac and the Equator Principles

In 2003, the Westpac Banking Corporation released an icy advertisement celebrating its identity as the 'first Australian Bank to sign the Equator Principles agreeing not to fund projects that endanger communities or the environment'.[7] Created by advertising agency The Campaign Palace – 'responsible for much of the advertising that has become part of Australia's popular culture' (McDonough & Egolf 2015, p. 252) – the advertisement features a single Adélie penguin atop an iceberg, with explanatory text and the company logo in the bottom right-hand corner. In 2008, the same advertisement was reprised to coincide with the Australian Emissions Trading Scheme (Nielson 2008) and Westpac's Climate Change Position Statement (Westpac Banking Corporation 2010).[8] The repeated use of such imagery speaks to the success of the earlier campaign, which 'capitalized on environmentalism as an opportunity' by 'extolling the virtues of the "Equator Principles"' (Fletcher & Crawford 2013, p. 208) – a voluntary risk management framework adopted by financial institutions 'for determining, assessing and managing environmental and social risk in projects' (Equator Principles Association, 2011). The advertisement thus presents Westpac as an environmentally friendly bank, with an active interest in reducing pollution. The same Adélie penguin image featured in the Westpac advertisement has made further appearances in the media, including as the introductory image to a special feature on climate change in *TIME* magazine (Kluger et al. 2007, pp. 54–55). This context further reinforces the environmental symbolism associated with this stock image animal celebrity.

The 2008 Westpac advertisement features the tagline, 'How many banks does it take to change a globe?'. This pun calls upon a history of jokes about changing light bulbs (also known as 'globes'), but also carries more serious undertones.[9] In this case, the globe in question is the Earth itself, a message reinforced by the answer provided – 'All of them' – and the change referred to is environmental. The fact that only a single penguin is depicted is also significant. Penguins are 'often considered the epitome of uniformity' (Leane & Pfennigwerth 2011, p. 37) and shown in large groups, preparing to dive into the water one after another. The use of a single penguin in this instance suggests that it takes one (animal or corporation) to dive in first, before others can follow. This idea of leadership is reinforced visually in the advertisement by the use of an iceberg as the second visual element in the scene. Icebergs bring to mind the 'tip of the iceberg' analogy, suggesting that only a small part of the company's environmental activities are outlined in the advertisement. As well as referring to the Equator Principles, the secondary advertisement text notes that, 'Westpac is now also the first Australian bank to sign the UN Global Compact CEO Water Mandate to tackle the emerging global water crisis'.[10] Reference to a water crisis makes the background image of an iceberg all the more pertinent – with two thirds of the world's fresh water locked up in glaciers and ice caps (United States Geological Survey 2016), the question of water as a resource is closely tied to any changes in that ice, such as melting.

The choice of an image drawn from the cryosphere for the Westpac advertisement is significant because it serves as shorthand for environmental change. In this case, the allusion is not to the observable melting of Antarctic ice, but to halting this process. In order to understand this message, the audience needs to be familiar with current discourse around climate change. Here, the message created by the advertisement is that Westpac is a responsible corporate citizen; by choosing to bank with them, customers are also choosing to help protect the environment – including the ice at the poles. This message was carefully considered by the advertising company. As Mark Sareff (2006, p. 25) from The Campaign Palace conceded to Westpac stakeholders, 'It's a fine line between building a brand for sustainability – and being accused of greenwash'. The challenge was to present the company as being socially responsible, without invoking a cynical response from Westpac's 'marketing-savvy consumers' (Sareff 2006, p. 25). The advertising company's solution was to use inference and association, and to employ Antarctic imagery to communicate a message about environmentally friendly behaviour.

Penguins serve a similar function in both the ABB and Westpac advertisements: standing on ice, they stand for Antarctica, which in turn stands for the environment as a whole. The birds are remarkably versatile, symbolically speaking – Leane and Pfennigwerth (2011, pp. 30, 32) have detailed how penguins have been 'adopted by environmental groups as poster children for remote wilderness regions' and anthropomorphised to 'stand in for humans without the added complication [in 'uninhabited' Antarctica] of actual human presence'. There are many other examples of advertisements that employ this technique. A series of advertisements about CFC-free aerosols which appeared in 1988 and 1989 featured a group of Adélie penguins under the heading, 'Meeting of leading environmentalists welcomes change in aerosols', to promote the role of new propellant technologies in protecting the ozone layer (Figure 8.2).[11] Here, the penguins are depicted as the ultimate environmentalists, as they live in the polar environment – a message that also comes through in the ABB and Westpac advertisements, where the birds are framed as having the most to gain from any change. Thanks to their upright bodies, 'tuxedo' feather patterns and comical waddle, penguins can also be used to suggest a human presence when there is none, giving the audience more reason to identify with the message. These examples illustrate the ways in which animals can become 'symbolic pawns in human debates' (Leane & Pfennigwerth 2011, p. 36) – including debates about climate, fragility and Antarctica.

Penguins aside, both the ABB and Westpac campaigns illustrate how, in light of the 'increasing global segment that values environmentalism' (Fletcher & Crawford 2013, p. 208), 'greening' a business is a lucrative proposition. In each case, the advertisement is earnest, suggesting that the consumers of ABB and Westpac products both care about environmental issues and are able to make sense of Antarctic imagery as shorthand for environmental issues at large. Antarctica is represented as the ultimate environment to protect, and this framing of the continent continues to be employed in advertising campaigns. As we

Figure 8.2 Penguin advertisement relating to CFC-free aerosols, 1988/1989
Source: The Aerosol Information Service, 'Happy 50th Anniversary'[12]

shall see, casting Antarctica as a fragile environmental treasure does not always lead to such solemn presentations as these examples. Indeed, the same imagery that is used to signal environmental themes has also been used to subvert the very narrative it has helped to build.

Melting ice: double takes and double meanings

Having gained sufficient traction as a vehicle for conveying climate narratives in the cultural sphere, the same imagery of ice and penguins became ripe for satire that consciously subverted the protection narrative. In early 2007, Italian clothing company Diesel launched a multi-platform advertising campaign entitled, 'Global Warming Ready'. Designed by the agency Marcel (Paris), the campaign rolled out newspaper, magazine, transit and outdoor billboard advertisements across the world (Diesel 2007a). Striking imagery 'depicting ordinary scenes in a surreal, post-Global Warming world' was at the heart of the advertisements (Diesel 2007a). Models were portrayed lounging in front of a range of well-known global landmarks, such as the Eiffel Tower in Paris, St Mark's Square in Venice and Mt Rushmore in South Dakota, but the scenes had been altered to suggest a much warmer climate: jungle in Paris, tropical birds in Venice, sandy beaches around Mt Rushmore – and penguins atop rocky outcrops rather than ice. Each image featured the same minimal text – a red Diesel logo in the bottom-right corner and a stamp in the top-right corner, proclaiming the models 'Global Warming Ready'. A press release from Diesel (2007a) described the campaign as 'a thought-provocative, international advertising concept designed to ignite debate while raising awareness of the issues surrounding climate change'.

Debate certainly ensued, with the relationship between climate change and consumption – specifically, the way the term 'global warming' was used a marketing tool – the focus of environmental criticism of the advertisements (Harrison 2007, n.p.). While Diesel claimed the campaign was an ironic way of provoking discussion at a time when, in the words of spokesperson Joelle Berdugo Adler, 'Global warming was a tremendously hot button issue' (quoted in Dahlen, Lange & Smith 2010, p. 157), the company was accused of being 'far less concerned with fomenting political activism and lifestyle change than they are with selling their brand' (Harrison 2007, n.p.). Pearse points out that, 'Diesel hasn't mentioned the [climate change] issue since the campaign and doesn't publish information on its own carbon footprint' (2012, p. 78), suggesting that the 'Global Warming Ready' campaign was little more than a lucrative way of capitalising on buzzwords. Nevertheless, the campaign was a success in terms of scale and critical acclaim – 'Global Warming Ready' went on to win a Silver Lion for Print at the 2007 Cannes Lions International Festival of Creativity (Ads of the World 2012).

While the power of advertising lies in its ability to persuade consumers to buy a product, the precise meaning of the 'Global Warming Ready' tagline remains ambiguous. Is it the scantily clad models themselves – and therefore the company that makes their scanty clothes – who are ready for a warmer climate? Or is the advertisement meant to suggest that the Diesel brand itself is 'going green' by taking an active stance in trying to combat climate change? Should the advertisements be read as a swipe at the concerns of environmentalists, a critique of commercial markets, a call to action or a playful amalgamation of all three? Supporting promotional materials add little clarity.[13] A short online film (Diesel 2007b) outlined, in a serious BBC-style, why global warming was 'bad', before announcing that, 'Global warming cannot stop our lives!' Images of the glamorous young people featured in the advertising campaign partying in exotic locations prompted Pearse to claim that, 'When it comes to the fashion industry and climate change, it *is* mostly about *looking* hot' (2012, p. 86). Taking a stance on global warming is typical behaviour for a brand that is trying to promote its environmental consciousness, but such a straightforward reading of the environmental message is denied by the presentation of self-centred people who appear quite content in their post-global warming lives.

Accompanying links and details on Diesel's campaign webpage are equally contradictory. At first glance, web materials appear to be irreverent: in order to save the planet, visitors to Diesel's website were exhorted to 'save the planet by having sex (quietly) to cut down on heating ... insulating homes with recycled denim, never taking a shower ... giving fashion magazines to grannies ... and getting rid of the fridge at home' (StopGlobalWarming.org 2007). At the same time, the site promotes Al Gore's famous 2006 climate change film, *An Inconvenient Truth*, and provides a link to Diesel's partner organisation, the online grassroots movement StopGlobalWarming.org. Diesel's statement on the movement's website suggested a modicum of self-reflection, with the acknowledgement that, 'We are only a fashion company and do not think that – with just one campaign – we can save the world' (StopGlobalWarming. org 2007). This statement could, however, be read as marketing spin that aims to preempt any critical media reaction to their campaign. It is not uncommon for advertising agencies to use dissonance and subversion to attract attention to brands and products marketed to young people (Andersson et al. 2004, p. 96); the shock factor and the pushing of boundaries are part of the attraction.

The 'Global Warming Ready' campaign may be designed to shock rather than to promote environmental values, but Diesel goes on to suggest that an advertising campaign can nonetheless have an impact:

> If our unconventional tone of voice and the reputation of our brand can grab and hold people's attention a little longer than a news feature can, make them think twice about the consequence of all our actions and realize our individual responsibility, then something at least will have been accomplished.
>
> (cited in StopGlobalWarming.org 2007)

The inclusion of such material in the campaign indicates an awareness of the scrutiny the campaign would – and did – provoke. Dahlen, Lange and Smith asked in 2010 (p. 157) whether Diesel's announcement that its products were 'Global Warming Ready' was 'a strategic shift to eco-fashion' or 'a short-term "green wash" use of tactical positioning to gain attention'. A third possibility exists – that the 'Global Warming Ready' campaign was a play on both, designed to highlight the issues around using buzzwords in adverting, while at the same time doing just that, by using the term 'global warming' in order to play to the consumer market. Diesel's clientele are mostly young, so have been exposed to the climate change message for much of their lives. Here, Diesel relies on that audience experiencing climate change exhaustion, and enjoying the naughtiness of revelling, however briefly, in the prospect of a warmer world. The campaign's intentional ambiguities, and the assumed sophistication of the target audience, provide important contexts for reading the Antarctic advertisement from the 'Global Warming Ready' series.

In the Diesel advertisement, then, existing visual language is employed to communicate ambivalent and subversive environmental values. The fact that an Antarctic setting – signaled only by penguins – was included in this series of

'Global Warming Ready' advertisements confirms that the continent has come to be understood as shorthand for environmental change. In this Diesel campaign, the melting of Antarctic ice is taken to its extreme conclusion, where there is no longer any ice left to melt. Despite calling upon the rhetoric of melt and change, the advertisement does not carry a sincere message in reference to Antarctica. On first glance, the advertisement seems to assume that, for its target audience, a balmy climate where one can wear a bikini is preferable to an icescape, and that the loss of the cryosphere is no great problem. Alternatively, the advertisement can be read as a reaction to the over-saturation of the media by narratives of climate change in which Antarctic imagery is subverted in order to play into the tongue-in-cheek nature of the overall campaign, thus gaining publicity for the brand at large. Whether one reads it as an ironic comment on how slogans can be used to sell, or as a denial of a bleaker future, the advertisement appeals to different values around Antarctica from those of the earlier ABB and Westpac examples. Antarctica is used as a tool through which environmental messages are propagated, represented, contested and recast for a range of different purposes.

Final reflections on a fragile continent

When it comes to representations of Antarctica, the frame of fragility is recurrent, and manifests itself in different ways. The case studies in this chapter have shown that varying connotations of ice are linked to cultural norms. These advertisements are all anthropocentric, presenting humans as the active agents – even when penguins are used as stand-ins for humans, as demonstrated in Westpac's 'Equator Principles' campaign. As such, they do not tell us about Antarctica as a place, so much as what Antarctica is used for – the continent becomes a symbol for environmental change. Imagery ranging from glaciers to penguins to the continent itself has been used to promote 'green' products, with a variety of companies capitalising on the idea that Antarctica needs to be protected. Growing consumer awareness of environmental issues from the 1980s onwards meant green initiatives became more valuable. So, too, did icy imagery. The rhetoric of climate change (often symbolised by Antarctica) saturated the media, including appearing in advertisements for companies that wanted to push their own 'green' credentials. Penguins and icebergs were employed to stand in for an environment that each company in turn could claim to be protecting, and the idea of Antarctica as a fragile place that is under threat entered into popular consciousness. Once polar imagery gained sufficient cultural traction to be easily understood as a symbol for climate change and the environment, the same imagery could be employed in subversive ways, as seen in the Diesel example.

In its fragility, Antarctica also makes humans fragile and vulnerable to the impacts of anthropogenic climate change. To take one dominant, and very visual, example, if Antarctica is at risk of melting, coastlines the world over are also at risk of flooding due to sea-level rise. This kind of bad-news story is not employed in our advertising examples, however; their purpose is to sell products or services to people, not to preach dystopian futures. This drive to sell lies behind the rhetoric of protection that urges consumers to buy this product and protect that ice. Such a message brings security and a feeling of agency to the consumers who make market choices based on their own environmental values, and brings profits to the companies who employ environmental rhetoric in their advertising campaigns. For Antarctica itself, such shifts in framing might ultimately make little difference. Back home, however, they reveal dominant values, and showcase what it is that is valued. No matter the tone or intended message, in the advertisements considered, Antarctica is there for the melting.

Notes

1 Andrea Juan's *Antarctica Project* (2005–2014) explores the collapsing ice shelves; Satoshi Itasaka's resin vase, entitled 'Larsen C', refers to global warming, with the artist stating that his artwork reflects that 'its disintegration, too, is just a matter of time' (Itasaka 2018).
2 When twelve percent of the Larsen C ice shelf broke off in July 2017, this event was also heavily covered by the media. It was framed as a climate change event, despite scientific statements to the contrary (see Luckman 2017; and Goodell 2017).
3 For a discussion of the term 'cryoscape', see Nüsser & Baghel 2014, p. 138.
4 For a data set of over 300 Antarctic advertisements, see Nielsen 2017 (Appendix).
5 ABB is a Swiss multinational corporation, founded in 1988, with headquarters in Zürich.
6 When it comes to polar scenes, penguins are often used as markers for the south, while polar bears indicate a northern setting. In the advertisement discussed here, a non-Antarctic species is utilised presumably (and ironically) because it offers a better 'Antarctic aesthetic' than the Adélies used in its previous iteration. King penguins are more colourful than the Gentoo, Adélie or Chinstrap varieties, and closely resemble the famous Emperor penguins.
7 This initial 2003 advertisement was part of the wider 'Building better lives for all Australians' campaign, run by The Campaign Palace (Sareff 2006).
8 The campaign was rolled out 'across a range of media, including television, newspapers and outdoor advertising' (Westpac Banking Corporation 2008).
9 Light bulbs themselves have a large environmental impact: Pearse (2012, p. 61) notes that 'worldwide, lighting generates around 1.9 billion tons of carbon dioxide each year – about three times as much as commercial aviation'.
10 The CEO Water Mandate was established by the United Nations Global Compact (2007) to mobilise 'business leaders for water stewardship'.
11 The advertisements, designed by Sydney-based agency Curtis Jones and Brown, originally appeared in the pages of women's magazines *Cleo, Cosmopolitan*, the *Australian Women's Weekly*, and also *Reader's Digest*, between November 1988 and February 1989. The use of penguins in this instance was particularly apposite, as the ozone 'hole' which appears annually above the Antarctic was first discovered by scientists making measurements at the British Faraday and Halley Stations on the Antarctic Peninsula (British Antarctic Survey 2015).
12 This advertisement, and several others examined in this chapter, can be accessed using the URLs provided in the References, as I was not granted permission to reproduce the figures in this forum. Note that, as websites are often ephemeral, the Internet Archive is a useful tool for accessing dated versions of each site. See <http://archive.org/web/web.php>.
13 These included the Diesel website, a documentary-style filmed advertisement, online links and endorsements. A booklet of the 'World's Coolest Hotspots Guide' was also available in store, and presented maps of various regions of the world affected by sea-level rise (including a green-coloured Antarctica).

References

ABB 2005, 'Cut 68 million tons of CO_2 and its (sic) amazing what you save', advertisement, *Animals in advertising – Penguins*, viewed 8 February 2019, <http://www. elve.net/panim/ill400/im411.jpg>.

ABB 2008, 'Cut 100 million tons of CO_2 and its (sic) amazing what you save', advertisement, *European Career Fair*, MIT European Club, p. 50, viewed 22 August 2017, <http://documents.mx/documents/2008-ecf-booklet-55a236eed0863.html>. Image also available at *Europolitics*, 13 March 2007, no. 3266, viewed 10 March 2019, <http://www.library.coleurop.pl/intranet/documents/eis/ep/ep3266.pdf>.

Ads of the World 2012, *Diesel: Global Warming North Pole*, WayBackMachine, viewed 18 December 2018, <https://www.adsoftheworld.com/cannes_lions_2007_winners_p ress_silver>.

Andersson, S, Hedelin, A, Nilsson, A & Welander, C 2004, 'Violent advertising in fashion marketing', *Journal of Fashion Marketing and Management*, vol. 8, no. 1, pp. 96–112.

British Antarctic Survey 2015, *The ozone hole*, viewed 22 August 2017, <http://www.ba s.ac.uk/data/our-data/publication/the-ozone-layer>.

Carvalho, A 2010, 'Reporting the climate change crisis', in S Allan (ed.), *The Routledge companion to news and journalism studies*, Routledge, Oxford, pp. 485–495.

Christensen, M, Nilsson, AE & Wormbs, N 2013, *Media and the politics of Arctic climate change*, Palgrave Macmillan, New York.

Dahlen, M, Lange, F & Smith, T 2010, *Marketing communications: a brand narrative approach*, John Wiley & Sons, Chichester.

Diesel 2007a, *Diesel launches global warming ready campaign*, Newswire (Canada) press release, viewed 22 August 2017, <https://www.newswire.ca/news-releases/diesel-la unches-global-warming-ready-campaign-533357341.html>.

Diesel 2007b, *Global warming ready*, online video, viewed 10 May 2016, <http://www. youtube.com/watch?v=V5OP3GMWWnY>.

Emmerich, R (dir) 2004, *The day after tomorrow*, motion picture, 20th Century Fox, Los Angeles.

Equator Principles Association 2011, *About the Equator Principles*, viewed 18 December 2018, <https://equator-principles.com/about/>.

Express Web Desk 2017, 'UAE firm plans to haul iceberg from Antarctica', *The Indian Express*, 7 May, viewed 10 May 2018, <http://indianexpress.com/article/world/to-ta ckle-water-scarcity-uae-firm-plans-to-haul-icebergs-from-antarctica-report-4644350/>.

Fletcher, R & Crawford, H 2013, *International marketing: an Asia-Pacific perspective*, Pearson Australia, Sydney.

Goodell, J 2017, 'The Larsen C crack-up on Antarctica: why it matters', *RollingStone*, 12 July, viewed 15 August 2017, <https://www.rollingstone.com/politics/features/ the-larsen-c-crack-up-in-antarctica-why-it-matters-w491929>.

Harrison, P 2007, 'Diesel ads hardly "Global Warming Ready"', *The Varsity*, 27 February 2007 (updated 11 January 2012), viewed 5 May 2014, <https://www.theva rsity.ca/2007/02/27/diesel-ads-hardly-global-warming-ready>.

Hicks, R 2005, 'Sustainability: can advertising save the world?', *Campaign*, 15 July, viewed 12 May 2016, <https://www.campaignlive.co.uk/article/sustainability-adver tising-save-world/486529#>.

Itasaka, S 2018, 'Larsen table holder', *Airgora*, viewed 18 December 2018, <https:// www.airgora.com/products/43>.

Juan, A 2005–2014, *Antarctic Project 2005–2014*, viewed 17 February 2014, <www. andreajuan.net/tpl/main.html#/app/proj/antart>.

Kluger, J, Baker, A, Bjerklie, D, Graham-Silverman, A, Sayre, C, Cray, D & Walsh, B 2007, 'What now?', *TIME*, vol. 169, no. 15, pp. 50–61.

Leane, E & Pfennigwerth, S 2011, 'Marching on thin ice: the politics of penguin films', in C Freeman, E Leane & Y Watt (eds), *Considering animals: contemporary case studies in human-animal relations*, Ashgate, Farnham, pp. 29–40.

Lindsey, R 2002, 'World of change: collapse of the Larsen-B ice shelf', *NASA Earth Observatory*, viewed 2 March 2016, <https://www.earthobservatory.nasa.gov/Fea tures/WorldOfChange/larsenb.php>.

Luckman, A 2017, 'I've studied Larsen C and its giant iceberg for years and it's not a simple story of climate change', *The Conversation*, 12 July, viewed 12 December 2018, <https://theconversation.com/ive-studied-larsen-c-and-its-giant-iceberg-for-yea rs-its-not-a-simple-story-of-climate-change-80529>.

Mayell, H 2004, 'As consumerism spreads, Earth suffers, study says', *National Geographic News*, 12 January, viewed 23 August 2017, <https://www.nationalgeographic.com/ environment/2004/01/consumerism-earth-suffers/>.

McDonough, J & Egolf, K 2015, *The advertising age encyclopedia of advertising*, Routledge, Abingdon.

NASA 2013, 'NASA scientists react to 400 ppm carbon milestone', *NASA global climate change: vital signs of the planet*, viewed 22 August 2017, <https://www.climate.nasa. gov/400ppmquotes>.

Nielsen, H 2017, 'Brand Antarctica: selling representations of the south from the "Heroic Era" to the present', unpublished doctoral thesis, University of Tasmania.

Nielson, L 2008, 'Emissions – who is trading what?', *Parliament of Australia*, viewed 24 August 2017, <https://www.aph.gov.au/About_Parliament/Parliamentary_Departm ents/Parliamentary_Library/pubs/BN/0809/Emissions>.

National Oceanic and Atmospheric Administration (NOAA) 2015, 'Greenhouse gas benchmark reached', *NOAA Research News*, 6 May 2015, viewed 13 December 2018, <https://research.noaa.gov/article/ArtMID/587/ArticleID/780/Greenhouse-ga s-benchmark-reached->.

Nüsser, M & Baghel, R 2014, 'The emergence of the cryoscape: contested narratives of Himalayan glacier dynamics and climate change' in B Schuler (ed.), *Environmental and climate change in South and Southeast Asia: how are local cultures coping?*, Brill, Boston, pp. 138–157.

Pearse, G 2012, *Green wash: big brands and carbon scams*, Black Inc., Melbourne.

Plester, J 2016, 'Continuing collapse of Antarctic ice shelves will affect us all', *The Guardian Online*, 16 May, viewed 25 May 2017, <https://www.theguardian.com/ news/2016/may/15/weatherwatch-antarctic-ice-shelf-collapse-global-implications>.

Rodger, A 2013, 'Antarctica: a global change perspective' in D Walton (ed.), *Antarctica: global science from a frozen continent*, Cambridge University Press, Cambridge, pp. 301–324.

Sareff, M 2006, 'Campaign for change', *Westpac Stakeholder Impact 2006*, Westpac Bank, pp. 26–27, viewed 20 December 2018, <https://www.westpac.com.au/docs/pdf/a w/SIR06_Customers.pdf>.

Savolainen, A 2004, 'Driving towards a better future', *ABB Review*, no. 4, pp. 34–38.

Secretariat of the Antarctic Treaty 1991, 'The Protocol on Environmental Protection to the Antarctic Treaty', 4 October 1991 (in force 15 January 1998). Madrid. 30 ILM 1461, viewed 29 August 2017, <https://www.ats.aq/e/ep.htm>.

Shackleton Foundation 2014, *Welcome to the Shackleton Foundation*, viewed 23 July 2014, <http://www.shackletonfoundation.org>.

Simpson-Housley, P 1992, *Antarctica: exploration, perception and metaphor*, Routledge, London.

Sörlin, S 2015, 'Cryo-history: ice and the emerging Arctic humanities' in B Evengård, J Nymand Larsen & Ø Paasche (eds), *The new Arctic*, Springer, Heidelberg and New York, pp. 327–339.

StopGlobalWarming.org 2007, *Featured partners: Diesel*, viewed 21 July 2015, <http://www.stopglobalwarming.org>.

United Nations Global Compact 2007, *The CEO water mandate*, viewed 20 March 2016, <https://www.ceowatermandate.org/what-we-do/faq>.

United States Geological Survey 2016, *Ice, snow and glaciers*, viewed 8 February 2019, <https://water.usgs.gov/edu/watercycleice.html>.

Vaughan, DG, Comiso, JC, Allison, I, Carrasco, J, Kaser, G, Kwok, R, Mote, P, Murray, T, Paul, F, Ren, J, Rignot, E, Solomina, O, Steffen, K & Zhang, T 2013, 'Observations: cryosphere' in TF Stocker, D Qin, G-K Plattner, M Tignor, SK Allen, J Boschung, A Nauels, Y Xia, V Bex & PM Midgley (eds) 2013, *Climate change 2013: the physical science basis. Contribution of working group I to the fifth assessment report of the intergovernmental panel on climate change*, Cambridge University Press, Cambridge, pp. 317–382.

West, D, Ford, J & Ibrahim, E 2012, *Strategic marketing: creating competitive advantage*, 2nd edn, Oxford University Press, Oxford.

Westpac Banking Corporation 2008, *PACT: Sustainability and Community News*, no. 5, viewed 20 December 2018, <https://www.westpac.com.au/docs/pdf/aw/pactV5.pdf>.

Westpac Banking Corporation 2010, *Westpac climate change position statement: financing the transition to a low carbon economy*, viewed 20 December 2018, <https://www.westpac.com.au/docs/pdf/aw/Transit_to_low_carbon_econo1.pdf>.

Williamson, J 2010, 'Unfreezing the truth: knowledge and denial in climate change imagery' in S Jerram & D McKinnon (eds), *Now future: dialogues with tomorrow 2010 series*, viewed 24 June 2014, <http://www.dialogues.org.nz>.

Yusoff, K 2005, 'Visualizing Antarctica as a place in time: from the geological sublime to "real time"', *Space and Culture*, vol. 8, no. 4, pp. 381–398.

Yusoff, K & Gabrys, J 2011, 'Climate change and the imagination', *WIREs Climate Change*, vol. 2, pp. 516–534.

Zazzle n.d., *Larsen B ice shelf collapse (picture Earth) shirt*, viewed 14 December 2009, <https://www.zazzle.com/larsen_b_ice_shelf_collapse_picture_earth_shirt-235664450722696309>.

Part 3
Inhabitations and place

9 Indigenising the heroic era of Antarctic exploration

Ben Maddison

In September 1899, an article in London's *The Strand Magazine* described the landing in Antarctica of the British Antarctic Expedition the previous February. The Anglo-Norwegian expedition leader, Carsten Borchgrevink, had chosen Cape Adare, the eastern headland of the Ross Sea, as the site at which the expedition would establish itself. The objective of the expedition was to be the first to explore the continental interior and to spend a winter living on the Antarctic continent, rather than aboard ship as Adrien de Gerlache's Belgian Antarctic Expedition in 1898 had done. Unloading the tons of supplies and equipment, which included a prefabricated living hut, ten tons of coal and – indicative of how little was then known about Antarctica – a large-bore gun suitable for killing polar bears, was an arduous task, as everything had to be rowed from ship to shore in whaleboats (Crawford 1998, pp. 84–85). As we shall see, divergences in the re-telling of this episode at the start of the so-called 'Heroic' Era of Antarctic exploration illustrate some of the fault lines around indigeneity with which this chapter is concerned.

On the afternoon of the second day of unloading, a hurricane swept down and the temperature fell to minus eighteen degrees Celsius. The *Strand* article described this episode in some detail, noting that

> Four members of the staff … were on shore and could not reach the ship. The only shelter was the tent, which they were obliged to bury with stones and to lash with ropes to prevent it blowing away … The hair of the party froze into solid lumps and the ice on their beards took hours to melt, while their clothes clashed with ice like coats of mail … A more awful experience it would be hard to imagine.
>
> (Newnes 1899, p. 284)

Borchgrevink's published account of the expedition described the episode in strikingly different terms. According to him, the only members of the expedition left on shore were two 'Lapps' or 'Finns': their names were Persen Savio and Ole Must, and they were Sami – indigenous people from northern Scandinavia (see Figure 9.1).

Figure 9.1 Persen Savio and Ole Must
Source: Scott Polar Research Institute, P75/33/1/32

It is unclear why Borchgrevink did not mention the scientists in his account of the episode, but several reasons help explain why the Sami were omitted from the version printed in *The Strand Magazine*. The article referred only to 'the staff', a term that in the contemporary vocabulary of Antarctic exploration referred to the scientific members of the expedition. Their exploits were likely to have been of most interest to readers of *The Strand*, a publication dedicated to extolling the achievements of the British middle class, most famously in the serialisation of Arthur Conan Doyle's Sherlock Holmes stories. By omitting the Sami from the episode, the survival of the 'staff' could be portrayed as a pure white hero story, a befitting start to the Heroic Era. To this end, the article commented blithely that 'the only shelter was the tent', leaving readers with the impression that this was just an ordinary tent, part of the general equipment of the expedition. But it was not any old tent, and it was not part of the general equipment of the expedition. It was the tent that the two Sami had brought with them from their homeland. They were 'undoubtedly the most experienced of the party at living under extreme [polar] conditions' (Crawford 1998, p. 43). In all probability, then, they had taken their tent on shore with them as a standard part of their polar equipment. If they had not done so, it is likely that the party would not have survived the night. The tent provided shelter from the hurricane, and its traditional 'wigwam' design, with a roof opening, allowed a fire to be built inside (Crawford 1998, p. 85).

In the early twenty-first century, it is hardly a profound insight to observe that indigenous presence was erased from accounts of colonial endeavours such as this. For decades, scholars of whiteness and indigeneity have been pointing out the ubiquity of such erasures, and their necessity in creating white-supremacist representations of colonial exploration and expansion as the autonomous activity and achievement of the European middle class. One of the consequences of what I call the 'historiographical lag' that has bedevilled Antarctic history is that scholars have only very recently started to apply these aspects of critical theory to Antarctic studies. Peder Roberts' essay 'The white (supremacist) continent' and Lize-Marie van der Watt's 'The whiteness of Antarctica' (Roberts 2016; van der Watt 2016) are two notable examples. Yet, despite this welcome focus on whiteness, very little sustained attention has been given to matters of indigeneity itself. Lynnette Russell's study of the role of indigenous people as sealers and whalers in the Southern Ocean, and Brigid Hains' work drawing parallels between the Australian outback colonial processes and Antarctic exploration are partial exceptions to this general rule (Russell 2012, 2018; Hains 2002). Nowhere is this lacuna more striking than in the historiography of the 'Heroic Era', where, despite a new plurality of historical actors – (most notably, expanding our understanding of the part played by women, workers and scientists) – there continues to be an embedded, and largely hidden, Eurocentrism. The assumption has been made that, because, as it turned out, there were no indigenous 'Antarcticans', and because Antarctic exploration was overwhelmingly undertaken by Europeans, indigeneity could not be relevant to understanding its history.

The intrinsic flaw in this assumption is illustrated by observing that relationships between indigenous peoples and Antarctica long pre-dated the explorations of the 'Heroic Era'. Rarotongan (Cook Islands) traditional knowledge maintains that a Polynesian navigator in the seventh century led a voyage far into the south, and reported encounters with icebergs and seals (McCann 2018, p. 87). Polynesian voyagers might have carried this traditional knowledge with them when, six centuries later, they discovered and settled in Aotearoa/New Zealand. Shortly after their arrival (c. 1270 CE) – and perhaps inspired by their ancestors' example – a small group voyaged even further south, settling for a short period in the subantarctic Auckland Islands. While this was not Antarctica itself, it nonetheless suggests a strong orientation to the austral pole, a relationship that is reflected in Māori interpretations of the aurora australis as the flickering flames of their ancestors' southern polar campfires (Maddison 2018a, pp. 72–73). Tasmanian Aboriginal peoples also established a relationship with the deep south via the aurora. The new Australian Antarctic icebreaker, set to replace the *Aurora Australis* in 2020, is named *Nuyina*, the *palawa kani* name for the southern lights, in acknowledgement of 'the interwoven history of Aboriginal people and the great southern land – Antarctica' (Australian Antarctic Division n.d.). In a similar move, a Māori *pou* [carving], titled the *Navigator of the Heavens*, was erected at New Zealand's Scott Base in McMurdo Sound in 2013, to acknowledge the strong historical links between southern Māori and Antarctica (McKerrow 2013). Indigenous connections with Antarctica continued in the nineteenth century, when Māori and Australian Aboriginal peoples worked in the sealing and whaling industries (Maddison 2014, pp. 76–77; Russell 2018).

Closer investigation of these instances is beyond the scope of this chapter, but it is nonetheless important to recognise that the intersection of indigeneity with Antarctica was neither initiated by – nor simply an artefact of – the forces that produced the Heroic Era. The longevity of that intersection notwithstanding, the early twentieth-century iterations of indigeneity and Antarctica were inextricably bound up with colonialism in ways that this chapter will explicate. Recovering these intersections is not, however, a straightforward exercise in empirical history. Identifying the historical agency of subaltern groups across the colonial world is notoriously hampered by their relative silence in the historical record (O'Hanlon 2012, pp. 73–74). This is particularly the case with indigenous peoples, and even more so when it comes to understanding their roles in Antarctic history. With the exception of one account mentioned below, indigenous voices and perspectives on Antarctica are notably absent from the written historical record. We thus have little direct evidence of how indigenous peoples experienced or related to Antarctic exploration during the Heroic Era. This limitation can be partly addressed by 'reading against the grain' – working interpretatively within the interstices, occlusions and inconsistencies of the dominant Antarctic narratives of the early twentieth century – to reveal historical presences and significances in ways that their writers had not intended.

This way of reading canonical texts of the Heroic Era is fundamental to the approach taken in this chapter. The argument also combines 1) the now generally accepted scholarly recognition that indigeneity was explicitly or implicitly involved in all colonial endeavours, with 2) the proposition that Antarctic exploration in the Heroic Era was inextricably an outgrowth of colonialism. By melding these approaches, and taking a trajectory from the empirical to the ideological and cultural, the chapter identifies several of the most important ways in which indigeneity and Antarctic history intersected in the Heroic Era. A conclusion reflects on some of the wider interpretative implications that flow from indigenising Antarctic history.

Traditional knowledge and Antarctic exploration

At its most explicit and empirical, indigeneity was present in the form of personnel on two, or possibly three, Heroic Era expeditions. As we have seen, there were two Sami on Borchgrevink's expedition. There might also have been an indigenous member of Scott's *Terra Nova* Expedition in 1911: Dimitri Gerov, a dog-handler on the expedition, was recruited from Siberia. More definitely, the Japanese Antarctic Expedition of 1910–1912, led by Lieutenant Nobu Shirase, employed two indigenous members. Yasanosuke Yamabe and Shinkichi Hanamori were Ainu from the subarctic Japanese province of Karafuto, now the Russian island of Sakhalin (Figure 9.2). Yasanosuke Yamabe was unique in leaving the only known indigenous account of Antarctic exploration in the Heroic Era.[1] The Shirase expedition was also unique because it broke with the European-dominated character of Heroic Era Antarctic exploration. Although this appears to erode the idea that Heroic Era exploration was conducted in a 'white supremacist' frame of reference, the positioning of indigeneity within the expedition echoed the European experience. As I have pointed out elsewhere (Maddison 2014, pp. 149–151), engaging in Antarctic exploration in the early twentieth century was a demonstration of the vigour of an up-and-coming colonial power such as Japan. Regardless of national origin, colonial powers – almost by definition, or as part of their inner logic – adopted very similar master/subaltern positionings of indigeneity. It is no surprise, then, that as an aspirant colonial power Japan replicated this pattern.

Almost as soon as the expedition landed in Antarctica, Yamabe and Hanamori's traditional knowledge and equipment came into play in an episode with striking parallels to that involving the Sami, Savio and Must, at Cape Adare. The official account describes how they

Figure 9.2 Shinkichi Hanamori and Yasanosuke Yamabe, wearing Karafuto dog fur suits, and holding Karafuto dogs. First published in 1910 in the magazine *Tanken Sekai*.
Source: Image provided courtesy of Hilary Shibata and the Shirase Nankyoku Tanjentai Kinenkan [Shirase Antarctic Expedition Museum], Nikaho City, Akita Prefecture, Japan

had intended to make a start at putting up the tent ... but what with the gale-force winds and the speed and strength of the tide ... decided to call a halt for the day, and all retreated to rest below deck. However the two Ainu, Yamabe and Hanamori, whose job it was to look after the sledge dogs, decided to camp out near where we had made a depot of our supplies.

The next morning, after the storm had abated, the landing party found that, 'The Ainu were both asleep and dreaming happily, wrapped up in their sleeping bags on mats spread over the floor of an Ainu-style structure made by putting together four bamboo poles and stretching a piece of hessian over the windward side' (Shirase Antarctic Expedition Supporters' Association [SAESA] 2011, pp. 150–152).

As this account indicates, the principal role of the Ainu revolved around the dog sledges. Their work involved caring for the dogs, organising the teams, packing the loads onto the sledges and driving the sledges. All this was undertaken using Ainu traditional knowledge and techniques – the arrangement of the dogs on the lines and the use of a short ski on one foot to assist steering were particularly important aspects. Sledging was the main form of transport used to explore into the interior on this expedition, and although occasionally other members of the expedition were allowed to drive the sledge, this was only when the two Ainu felt that their skills would be better used for navigating and way-finding (SAESA 2011, p. 175).

Once the sledging had begun, their expertise came to the fore. On the morning of the first day of sledging

> The two Ainu were so eager to get started ... that as soon as they had finished breakfast they set to work, knowing that there was not a moment to lose. As this was the first time they had tried to load all our many supplies and equipment onto the sledges this took some time. To make matters more difficult it was so cold they could not remove their gloves for even a second, and it is a very difficult thing to load a sledge wearing thick gloves. Despite all such obstacles the amount of work these two men did would have brought the devil himself to his knees with exhaustion, and soon enough everything was going as planned.
>
> (SAESA 2011, pp.157–158)

The two Ainu brought a very high level of sledging skill into Antarctica. As Hilary Shibata has demonstrated, the Ainu 'set a record for the fastest dog sledge travel in Antarctic history' (Shibata 2018; pers. comm., 2 February 2019). Despite their expertise and energy, a range of other tasks performed by the two Ainu demonstrated that they were positioned at the bottom of the expedition's racial hierarchy. On the voyage out to Antarctica they not only worked hard day and night looking after the dogs, but also worked alongside the sailors altering the set of the sails, took it in turns to do odd jobs for the scientific staff and assisted other members of the expedition in general.

(SAESA 2011, p. 105)

The two men were also responsible for much of the domestic labour of the expedition. Some of their main tasks were collecting and melting ice for bathing, cooking, seal hunting (SAESA 2011, p. 100) and, because of 'the skills they had acquired as a result of their long training in sea-mammal cuisine' (p. 121), the butchering and cooking of the meat. Similarly, killing and preparing a dolphin 'was entrusted to the two Ainu ... who had much experience of such matters' (p. 223).

Despite all this heavy labour, and the recognition by other expedition members of their contributions, Yamabe and Hanamori were nonetheless relegated to the periphery. Most graphically, they were marginal or non-existent figures in the traditional 'furthest South' and 'planting the flag' photographs published in the official record of the expedition. Of the sledging party, only the non-indigenous Miisho, Shirase and Takeda were shown, presumably with one of the Ainu operating the camera (SAESA 2011, pp. 172–173). The peripheral status of the Ainu was also evident in other ways. In some instances, Yamabe and Hanamori did not sleep in the main expedition tent, although whether this expressed their personal choice or an intentionally racialised division of space remains unclear (p. 393). When it became necessary to reduce loads and share sleeping bags, this division came into play even more strongly, with the indigenous men in one bag and the non-indigenous in another (p. 161).

The presence of two indigenous people served an additional purpose by adding exoticism and authenticity to the expedition. At a formal farewell celebration at which members of the Japanese Imperial family officiated, it was noted that, 'The two Ainu dog drivers were particularly resplendent due to their imposing height and build'. Thus differentiated by their stature, Yamabe and Hanamori were also differentiated by their response when the royal Countess gave each member of the expedition a cup of sake. Although 'everyone was moved by the warm hospitality of this gesture', this

was especially true of the two Ainu. ... They looked about, their big round eyes full of surprise and apprehension as the Countess herself filled their cups, and in the end were so overcome by their emotions that the tears flowed freely down their cheeks. Everyone present was deeply moved on seeing this, and some of us were obliged to make discreet use of our handkerchiefs.

(SAESA 2011, p. 44)

In similar vein, the presence of Yamabe and Hanamori, and their traditional knowledge, exoticised some of the key moments of the expedition.

Appropriately, the expedition's ship was greeted by twenty orca as it entered the Bay of Whales. The two Ainu began

> praying fervently to these whales, and when we asked them about this later were told that the killer whale was the emissary of the sea god and should therefore be accorded the same reverence. ... Assuring us that with the killer whale god protecting her, our *Kainan-maru* was sure of a safe voyage, the Ainu offered even more fervent prayers to the sea.
>
> (SAESA 2011, p. 129)

Yet, when all was said and done, the overall status of Yamabe and Hanamori was perhaps best illustrated when the expedition made its return landfall in Aotearoa/ New Zealand. While the other members of the expedition went ashore to celebrate, they remained on board, 'tidying up the decks' (SAESA 2011, p. 225).

Just as the Ainu were essential to the sledging operations and exploration achievements of the Shirase expedition, the Sami's prowess in polar travel provided the most important contribution to the Borchgrevink expedition's successes. As Louis Bernacchi, one of the expedition members put it,

Figure 9.3 Ole Must collecting penguin eggs at Ridley Beach, Cape Adare, Antarctica, November 1899

Source: LC Bernacchi photograph. LC Bernacchi Collection, Canterbury Museum, New Zealand

Savio and Must had greater learning and instinctive understanding of how
to live at intolerably low temperatures and to cross great ice-bound wastes
than anyone else on the expedition ... the success of any journey into the
interior could well depend upon the skills of the two Lapps.

(Bernacchi quoted in Crawford 1998, p. 125)

Drawing on this expertise, one or both of the Sami were taken on the expedition's various trips into the interior. Dog and sledging work was augmented
by tasks considered appropriate to the skills and status of the Sami, such as
hunting seals or, as seen in Figure 9.3, collecting penguin eggs for food.

Not surprisingly, their skills and knowledge notwithstanding, Savio and Must
'were treated politely but as inferiors' on the expedition (Bernacchi quoted in
Crawford 1998, p. 125). As with the Ainu, their roles combined polar expert,
labourer and exotic adornment. They occupied this latter role at the reception
given to the expedition on its arrival in Hobart, Tasmania. The *Mercury* newspaper reported that, 'As the invited entered they passed between a pair of Finns
in the picturesque costume of their country' (quoted in Crawford 1998, p. 59).
Occasionally, however, this kind of Orientalism could backfire, resulting in a
momentary carnivalesque overturning of the social order. Just before the
expedition's departure from Hobart for Cape Adare, there was what the expedition member Louis Bernacchi, described as 'A rather amusing episode'. 'One
of our Lapps', he wrote proprietorially, 'was accidentally taken to a reception at
Government House'. A guest, running into 'a gentleman in gorgeous uniform
(native dress) (sic) took him to be some distinguished officer' from the expedition ship. When this 'gentleman' arrived at Government House, the Gover-
nor 'knew him to be one of the Lapps in charge of the dogs, and so did most
of the guests. The former were (sic) of course horrified, and the guests were
intensely amused. It was a ludicrous situation' (Bernacchi quoted in Crawford
1998, p. 18). The Sami also added exotic authenticity to the most poignant
occasion of the expedition. Borchgrevink described how, at the burial of the
biologist Nicolai Hansen, 'The two Finns held a lappish [sic] service', and
noted that, 'It was touching to see these two simple-minded children of nature,
in their native costumes, standing with bared heads in the cold, singing and
talking alternately to Mr Hanson's dead body' (Borchgrevink 1980, p. 191).

Indigenous technologies

Even when indigenous people were not physically present, critical polar technology and skills gave them a proxy presence on most Heroic Era expeditions. Sledges were either made in Scandinavia – or in the case of the Japanese expedition, on
the island of Karafuto – by indigenous craftworkers or using techniques developed
by them over centuries (SAESA 2011, p. 280). The frames of sledges were made of
wood spars, lashed together with hide straps which allowed them to flex as they
crossed the rough Antarctic snow and ice fields. Even the capacities of sled dogs
were not simply a species capacity, but reflected the genetic impress of centuries of

indigenous polar mobility. Australian expedition leader Douglas Mawson (1996, p. 143) recognised that, 'The desire to pull is doubtless inborn, implanted in a long line of ancestors who have faithfully served the Esquimaux [sic]'. Skis made their first appearance in Antarctica on the expedition that marked the start of the Heroic Era, Borchgrevink's *Southern Cross* expedition. Although widespread in Scandinavia, skiing was little understood in many parts of the world in the early twentieth century. As Shirase commented in 1912, 'In future it will be important for explorers in snow-covered regions to study and perfect this new art of skiing' (SAESA 2011, p. 280).

New it might have been to the rest of the world, but to the Sami it was an activity with deep historical roots. Skiing had been invented in Scandinavia and northern China before 5000 BCE, and in the early twentieth century was widely used across the Scandinavian Arctic. Rock carvings from Norway dating from 4000 BCE show that the initial practice was to use a single stick as a ski-pole (Huntford 2009). The Sami were the bearers of this practice and, displaying this remarkable cultural and technological continuity, Savio and Must brought this ancient indigenous technique to Antarctica (Figure 9.4), where it was adopted by Borchgrevink and other members of the British Antarctic Expedition (Figure 9.5).

Figure 9.4 Persen Savio in traditional garb on skis, Cape Adare Antarctica 1898–1900
Source: LC Bernacchi photograph. LC Bernacchi Collection, Canterbury Museum, New Zealand. When this image was used in Borchgrevink's account of the expedition, it was revealingly captioned 'Savio travelling with Indian gear' (Borchgrevink 1901, p. 78)

FARTHEST SOUTH EVER REACHED
BY MAN.

Figure 9.5 Carsten Borchgrevink using Sami ski technique, c. 1900
Source: Borchgrevink, C 1901, First on the Antarctic continent, being an account of the
British Antarctic Expedition, 1898–1900, G Newnes Ltd, London, p. 295

To judge by images such as 'The Results of Ice Pressure from the South' in
Captain Scott's account, and Ernest Shackleton's 1902 photograph 'Practice on
Skis, Ross Island' (Figure 9.6), single-pole technique was also used on Scott's
1901–1904 *Discovery* expedition (Royal Geographical Society, Shackleton
Collection; Scott 1907, pp. 284–285). While Amundsen's expedition used two
ski poles, he nonetheless reckoned that skis 'played ... possibly the most
important of all roles, on our journey to the South Pole' (Amundsen 1912, p.
89). Fellow Norwegian Fritjof Nansen also acknowledged the indebtedness of
Antarctic exploration to indigenous technologies and their inventors, writing of
Amundsen's expedition that

> The means used are of immense antiquity, the same as were known to the
> nomads thousands of years ago ... it was all this, with the dog as a draught
> animal borrowed from the primitive races, that formed the foundation of
> the plan and rendered its execution possible.
>
> (Amundsen 1912, p. xxx)

Figure 9.6 'Practice on Skis, Ross Island', E. Shackleton, 1902
Source: Ernest Shackleton Collection, Ref. S0011524, ©Royal Geographical Society (with IBG)

Other indigenous technologies or discoveries also played an important part in Antarctic expeditions. *Finnesko* were boots made from reindeer fur.[2] Some appreciation of their importance in Antarctic exploration lies in Captain Scott's view that, 'Of all parts of the person of which it is necessary to have care the feet are the most important' (1905, vol. 1, pp. 396–397). Scott also gives a sense of the care that must have gone into the invention and development of the *finnesko*. 'The sole', he wrote, 'is made from the covering of the forehead [of the reindeer] both on account of the thickness of the pelt in this part, and also to obtain the twist in the growth of the hair which gives the boot a better chance of gripping on a slippery surface' – much the same as using skins on ski. Some *finnesko* were also made from the reindeer leg skin (pp. 292, 397).

Finnesko were first used in Antarctica by the Sami on the Borchgrevink expedition, and soon became standard kit for other Heroic Era expeditions. Members of the *Discovery* expedition quickly abandoned both 'English leather boots' and ski boots, in favour of *finnesko*, as they found them to be 'the most warm and comfortable foot-gear imaginable' (Scott 1905, vol. 1, p. 292). The special qualities of *finnesko* were maximised when they were used with sennegrass (*Carex vesicaria*), a dried sedge that was arranged in the boot. Mawson described them as 'the favourite foot-gear' on his expedition to East Antarctica in 1911–1913. They were, he explained, 'soft and commodious' and, 'Once these were stuffed with Lapp saennegras or manilla fibre, and the feet covered with several pairs of socks, cold could be despised unless one were stationary for some time or the socks or padding became damp' (Mawson 1996, p. 112). One of the advantages of using sennegrass was that, when the *finnesko* were not in use, any moisture that had accumulated on the grass froze and could be shaken off (Scott 1905, vol. 1, p. 397). Savio and Must transmitted this traditional knowledge and taught the Europeans how to use boots and sennegrass together. As Borchgrevink acknowledged, 'The great thing seems to be to arrange the grass properly in the boots, and although we all tried to imitate the Finns in their skill at this work, none of us felt as warm on our feet as when they had helped us' (1980, p. 257). Drawing on this experience, subsequent British expeditions took cases of *finnesko* and tea chests full of sennegrass (for a more detailed discussion see Maddison 2018b).

The same process of indigenous technology supplanting the non-indigenous was also evident on the Shirase expedition. Non-indigenous expedition members soon abandoned their European-style felt cavalry riding boots 'as they had proved far too heavy for such strenuous physical activity'. They adopted, instead, Japanese 'traditional straw snow-boots' known as *waragutsu* (SAESA 2011, p. 147 and fn 74, p. 390). Evidently these boots worked like *finnesko* – they were very light and 'did not feel cold even when the insides of the straw-boots became damp' (pp. 273–274). They also had the advantage of making it unnecessary to wear crampons in many conditions, even when 'carrying heavy loads on icy slopes or up steep snow hills' (pp. 273–274). However, *waragutsu* wore out quickly and needed to be replaced once or twice a day. The Ainu, on the other hand,

> 'wore sealskin boots, and these had all the advantages of the straw boots without suffering from any of their disadvantages. They were not only waterproof and light, but also helped to prevent one slipping on the ice.'
>
> (pp. 273–274)

Soon, 'all members of the [expedition] wore them and found them very useful indeed' (pp. 273–274).

Other types of traditional indigenous clothing were also used. Mawson's expedition used Arctic wolf-skin mittens, while Karafuto dog-skin mittens and jerkins were *de rigueur* on the Shirase expedition (SAESA 2011, p. 159). On the Amundsen expedition, the men's clothing was made from 'nearly two hundred and fifty good reindeer-skins, dressed [and presumably hunted] by the Lapps'. Once the skins had been procured, they were made up into polar clothing 'after the pattern of the Netchelli Eskimo'.[3] Each expedition member also 'had a suit of sealskin from Greenland' (Amundsen 1912, pp. 76–79). On Borchgrevink's expedition, Savio and Must 'were busily engaged in making fur suits out of seal skins', which, as Borchgrevink discovered, was 'the only clothing which keeps the cold out on windy days'. They also made up 'around fifty pairs' of *finnesko* on the spot (Borchgrevink 1980, p. 183). In a unique example of the incorporation of southern hemisphere indigenous products into Antarctic expeditions, members of Otto Nordenskjöld's Swedish Antarctic Expedition of 1901–1904 kept themselves warm using cloaks and sleeping bags that they had made from the skins of guanaco (related to llama[4]) that they procured from the Ona (indigenous peoples of Tierra del Fuego) when they stopped to undertake scientific work in Tierra del Fuego en route to Antarctica. Nordenskjöld described the luxurious furs as 'a most valuable addition to the ordinary outfit, for they weigh very little, but, thanks to them, one never, or seldom, feels the cold' (Nordenskjöld & Andersson1977, p. 217).

Indigenous inhabitants?

'Are there not people or unknown animals in the regions around the south pole?', asked Frederick Cook in 1900 (1980, p. 461). The American explorer, who had voyaged to Antarctica as the doctor on Adrien de Gerlache's 1897–1899 *Belgica* expedition, posed this rhetorical question towards the end of his account of the expedition. He went on to answer it in a circumspect fashion that left the issue open. 'From the explorations thus far', Cook wrote, 'we have no reason to hope for any startling discoveries of human or other animal life' (p. 461.) The Antarctic region was, in his view, 'far superior to the arctic' as a place for human existence, especially given the amount of easily procurable food (pp. 461–462). However, he concluded, 'No reliable traces of ... human beings have been found' – just 'one doubtful sign ... about fifty clay balls, perched on pillars' all of which, according to the discoverer, 'had the appearance of having been made by human hand' (pp. 461–462).[5]

Cook's discussion indicates that, even when indigenous people were not materially present in Antarctica, either as individuals or in their proxy-presence as technology, they were brought there in the minds of explorers. In 1520, Ferdinand Magellan and his party became the first Europeans to sail through the passage between the Atlantic and the Pacific Oceans, later called by the Spanish *Estrecho Magellanese* [Magellan Strait], erasing the names given to it by the Tehuelche, to the north, and the Ona, to the south. Drawing on prevailing geographical assumptions, it was believed the southern shores of the strait were the start of a great landmass which extended to the South Pole. As they sailed through the passage, Magellan's crew also observed people on the southern shores, probably members of the Ona nation. Putting these two observations together suggested the possibility of an inhabited southern continent, and, when the cartography of Magellan's voyage was incorporated into European mapping of the world, the idea of an Antarctic continent with an indigenous population came with it (Gurney 1998, pp. 8–13). This notion persisted for centuries. Even though Dutch, Spanish and English exploration in the late sixteenth and early seventeenth centuries corrected the Magellanic geographical error by showing that there was ocean, not land, to the south of Tierra del Fuego, the idea of an inhabited Antarctica was remodelled as European colonialism pressed further into the southern hemisphere. On the eve of Captain James Cook's watershed circumnavigation of Antarctica in 1775–1778, geographical pundits, such as Britain's Alexander Dalrymple, speculated about a southern supercontinent, possibly extending to the South Pole, with a population of some fifty million (Gurney 1998, pp. 16–17).

The exploration of Antarctica in the first half of the nineteenth century added greatly to Western knowledge of the continent, but being confined to ships, European explorers only nibbled at the coastal fringe, so the question of whether Antarctica was populated persisted. An American newspaper reported in 1820 that some recently discovered Antarctic islands were 'covered with snow, an abundance of seals and whales – no inhabitants' (quoted in Gurney 1998, p. 146). The idea of a populated Antarctica was disseminated in North America by pundits like John Symmes – who theorised a populated 'hollow earth', with the poles as access points – and works such as Edgar Allan Poe's *The Narrative of Arthur Gordon Pym of Nantucket*, which, when it was first published in 1838, was presented as an actual account of a whaling ship that was attacked by a 'native tribe' in the region of the South Pole (Maddison 2014, pp. 51–52).

When exploration resumed in the late 1890s, the question of whether Antarctica was inhabited still hovered unresolved in the consciousness of what would become known as the 'Heroic' generation. The plausibility of the idea was supported by works such as *The Voyage of Will Rogers to the South Pole,* which, like Poe's earlier work, was presented to the public as a true account of a man's experiences living with a tribe at the temperate South Pole (Spotswood 1888). These 'factual' accounts were buttressed by the many late-nineteenth-century novels that populated Antarctica by making their central premise the existence of people at the South Pole. The presence in popular culture of the idea that Antarctica might be inhabited was strikingly illustrated by an exchange that occurred on the return of Borchgrevink's British Antarctic Expedition, 1898–1900. The expedition's first homeward port of call was Lyttelton in Aotearoa/New Zealand, where Borchgrevink telegraphed his sponsors in London with news of the expedition's safe return. As Borchgrevink was waiting for the telegraph connection to England to be made, the New Zealand telegraph operator turned to him and reported that, 'Some very distant official [in London] desired to congratulate [him] on his safe return' and 'wished very much to know whether [he] had found people "down there"' (Borchgrevink 1980, pp. 294–295).

If the idea of an inhabited Antarctica lived on in the popular imagination, its shadow also lingered in the corners of the minds of at least some of those who explored the Antarctic continent. While most probably believed that Antarctica was uninhabited, nagging doubts persisted. One day, after their Antarctic winter in 1903, the Swedish explorer-scientists Otto Nordenskjöld and Ole Jonassen were sledging out from their hut at Snow Hill Island. In the distance they spied three totally unexpected figures and, in the moments of uncertainty about their identity as the two parties converged, the old questions about the uninhabited continent surfaced. Observing the party's unkempt appearance and blackened faces, Jonassen 'suggested that they might be members of some unknown Antarctic tribe'; he advised Nordenskjöld to 'take out [his] revolver in order to be prepared for all eventualities' (Nordenskjöld & Andersson 1977, p. 307). As the two parties moved closer, Nordenskjöld saw men 'who were as black as soot', displaying

> A mixture of civilization and barbarousness; my powers of guessing fail me when I endeavour to imagine to what race of men these creatures belong ... it was I who was civilized, and these men were the savages, reminding one of Australian aborigines, or some other low race of human beings!
>
> (pp. 307–310)

Amundsen's comments on this episode displayed the impress of racial hierarchisation on his outlook: the 'three figures' Nordenskjöld came across, wrote Amundsen, were 'doubtfully human' as they 'might at first sight have been taken for some of our African brethren straying thus far to the south' (Amundsen 1912, pp. 32–33). As it turned out, the 'Antarctic tribe' was three members of the expedition who had become separated from the main party for many months; they had covered their faces in seal blubber and soot to protect against sunburn. Their reunion and return to the main party was accompanied by a jocose parody about blackness, itself a common accompaniment to Antarctic expeditions. On another occasion, Nordenskjöld remarked that his sledging companions 'were half black in the face, and more like Indians that anything else' (Nordenskjöld & Andersson 1977, pp. 238–239). On the Shirase expedition, a similar response to surrogate indigeneity also prevailed: 'In the hot sun', the official account records,

> our faces were soon burnt as black as lacquer ... leaving only the whites of our eyes and our teeth in such stark contrast that we looked like sons of Africa! We laughed at each other's appearance, joking that there was no risk of getting lost in the snow looking like that!
>
> (SAESA 2011, p. 144)

Even where the sun did not inspire it, there were many other ways to deploy racial discourse. On Scott's *Discovery* expedition, the lead dog was called 'Nigger', a name that 'wholly failed to convey the grandeur of his nature' (Scott 1905, vol. 1, p. 408). Practically every dog team had its 'Nigger', and if you didn't have dogs, as with Scott's *Terra Nova* expedition, the cat would do (Cherry-Garrard 1965, p. 42). If there were no animals through which to deploy racial discourse, other opportunities invariably arose. Amundsen described how a celebration on his expedition, included

> a nigger in a tail-coat, a silk hat and a pair of wooden shoes. Black as he was, we saw at once that it was the second in command who had thus disguised himself. The mere sight of him was enough to set us all shrieking with laughter ... He was intensely amusing.
>
> (Amundsen 1912, p. 149)

Similar comfort in the dichotomous hierarchisation of black and white was garnered in the racial stereotypes deployed in the 'nigger minstrel performance' on Scott's *Discovery* expedition (Scott 1905, vol. 1, pp. 332–334). These ways of deploying racial discourse suggest something of the values and operation of racial hierarchies in Antarctic exploration that underpinned how contemporaries thought about Antarctic indigeneity.

An absent presence

In his book, *Antarctica: An Absent Presence*, sound artist Philip Samartzis describes Antarctica as a 'place marked by a strong sense of absence, empty and white, deserted and silent' (2016, p. 20). Although this description draws uncritically on standard tropes about the Antarctic landscape, we can adapt the useful phrase 'absent presence' to our discussion of indigeneity by including the human in its scope. This allows us to explore the ways in which indigeneity paradoxically 'appeared' in Antarctica through its own negation.

Admiral Byrd in the late 1920s was probably one of the last Antarctic explorers to hold onto the possibility that Antarctica was inhabited (Griffiths 2007, p. 243). Yet, even Heroic Age explorers who believed that Antarctica was uninhabited, found something unsettling and counter-intuitive about engaging with that rare entity, a real *terra nullius*. [6] Borchgrevink revealed much about this anxiety, as he mused on the episode which began this chapter. Recall that a party, stranded on shore by a storm, took shelter in the tent that was part of the polar kit of the two Sami men on the expedition. While the account published in *The Strand Magazine* erased the Sami, Borchgrevink's rendition made them the centre of the story. According to him, Savio and Must were 'the first [people] to sleep on the Antarctic continent'. He considered this to be a redemptive historical moment. It was 'curious', he wrote, that 'the two Finns were destined to be the first to sleep on shore', because this made them 'the first to inhabit that large southern continent', and as 'natives of the corresponding latitudes of the Northern hemisphere ... they, as children of nature from the north, in a way replaced the want of natives in the south' (Borchgrevink 1980, pp. 135, 232).

When Borchgrevink used the word 'want', he meant more than 'lack' or 'absence'; the word brought with it connotations of need, longing, desire, a yearning to have something wrong put right. What was wrong in Borchgrevink's mind – and which the two sleeping Sami put right – was the absence of indigenous inhabitants in Antarctica. The feeling of 'absence' or 'lack' to which Borchgrevink was so sensitive suggests a tendency to normalise indigenous presence as an accompaniment to the colonial endeavour. Indeed, such an awareness of this absence seems to have profoundly shaped the ways in which many explorers responded to their Antarctic experiences. Borchgrevink comments at one point that the 'centuries of heaped up solitude' in Antarctica shaped his experience of the landscape. In doing so, he was expressing both an existential and an historical loneliness generated by being in a place without a pre-existing population or human history. Captain Scott felt a similar sense of loss and absence in the 'vast ice-solitudes' and the 'frowning desolation of the hills' around McMurdo Sound in 1902. 'How strange it all seems! For countless ages', he wrote,

> great sombre mountains about us have loomed through the gloomy polar night with never an eye to mark their grandeur, and for countless ages the windswept snow has drifted over these great deserts with never a footprint to break its white surface.
>
> (Scott 1905, vol. 1, p. 308)

Frequent expressions of aloneness made by many other Antarctic explorers resonate with Scott's sentiments. The landscape around Snow Hill Island inspired Nordenskjöld to ruminate that the expedition had been but 'a brief moment – a slight track amidst the snow'. Once it had left, things returned to their prior state: 'the eternal world of ice' lay there 'as lonely as before. Once more the penguins live their quiet lives; and the storms drive on in paths unmarked by human eye' (Nordenskjöld & Andersson 1977, p. 600).

One of the most potent ways for explorers to assuage that sense of historical loneliness was to record the impact of their expeditions' presence on the landscape. Captain Scott, observing the activity at Hut Point, noted that, 'For one brief moment the eternal solitude is broken by a hive of human settlement' (1905, vol. 1, p. 308). Not too far away in space and time, Amundsen was deploying a similarly comforting set of resonances. At Framheim, the expedition base on the Ross Ice Shelf, 'The smoke rose gaily from the shining black chimney, and proclaimed that now the Barrier was really inhabited ... It is a little thing but nevertheless it means so much' (Amundsen 1912, p. 198). The Norwegian later confided to his diary that, 'It is strange indeed here to go outside in the evening and see the cosy, warm lamp-light through the window of our little snow-covered hut, and to feel that this is our snug, comfortable home on the formidable and dreaded Barrier' (p. 254). When members of the Shirase expedition found themselves 'amidst the vast and lonely wastes of Antarctica, thousands of miles from any other human being', they became 'anxious, lost, and enveloped in a thick white fog, and these eerie echoes [of whales blowing in an unknown bay in the ice edge] struck horror into our hearts and left us with a feeling of dreadful awe'. Even the calls of a rescue party took a while to assimilate – a voice 'faint and far away and sounding like a lost soul calling from the pit of hell' (SAESA 2011, p. 177). For them, the 'desolate landscape barely touched by man's presence since the very beginning of time' was made less threatening when it 'echoed with the sound of our snores' (p. 173). When the team reached its 'furthest south' on the Ross Ice Shelf, the expeditioners raised the Japanese flag 'in the midst of this vast desert of snow'. It 'fluttered on the polar breeze, its pure red reflected on the whiteness of ice and snow that had endured from time immemorial'. It was envisaged that this place, hitherto lost in time – a place of 'time immemorial', with no memorials or memories to anchor it in chronology – would in 'thousands nay tens of thousands of years from now, indeed for as long as this Earth shall last, prosper as the territory of Japan'. And, it was thought, 'Although we know it is an uninhabited continent, thousands of years from now the smoke of home fires will surely rise into the air, and there will be a whole town built here with carriages going to and fro' (p. 173).

Conclusions

Paying attention to indigenous personnel, indigenous technologies and the 'absent presence' of humans allows us to identify key points where indigeneity and Antarctica intersected in the Heroic Era. We can, however, go further than this, by considering the reflections made by French explorer Jules Dumont d'Urville when he claimed Adélie Land for France in 1840. It was, he said, a place where 'never before had man's voice rung out', a 'land which no human creature had either seen or stepped on before'. The French claim to Adélie Land, he emphasised, was 'incontestable', as it was 'a wholly pacific conquest' that 'dispossessed none' and had been undertaken with none of the 'abuse which has been born of such acts' in many other places (Dumont d'Urville 1987, vol. 1, pp. 476–478; vol. 2, pp. 471–473). In critiquing the violence of colonialism elsewhere, Dumont d'Urville was implicitly defining its absence as one of the unique historical characteristics of Antarctica.

Dumont d'Urville's remarks reflected a strand of thinking that had accompanied colonialism since at least the late sixteenth century, when the monk Las Casas debated the morality of colonialism and its intrinsic violence with Spanish court officials. Such concerns continued into the nineteenth and early twentieth centuries. Although never dominant, they articulated a moral doubt about colonialism, the bad conscience with which colonial societies had to grapple. These 'whisperings in the heart' served to cast a shadow over 'the sunny narratives' of heroic progress that so often was celebrated in the building of colonial societies (Reynolds 1998, pp. 245–246).

As Dumont d'Urville's comments suggest, one of the most important differences between the Antarctic and other instances of colonialism was the human absence that we have been discussing. Unacknowledged as it has been by historians, the 'want of natives in the south' has nonetheless had a profound impact on how Antarctica and its history has been seen. At the most immediate level, it allowed the exploration of the continent to be construed as ethically unambiguous in ways that were precluded in other sites of colonial domination. In turn, this allowed a whole host of meanings about human endeavour and possibility in Antarctica to flourish. Indeed, looking through the prism of indigeneity, we can see how this 'absent presence' was one of the enabling conditions for some of the most prominent tropes about Antarctic exploration that developed alongside the exploration of the continent. It is hard to imagine that the 'heroism' and the display of 'the highest qualities of human endeavour' that was supposed to be the contemporary – as well as, until recently, the historiographical – epitome of Antarctic exploration in this era would have been possible if its history had been marked by the bloody slaughter that characterised colonialism elsewhere. And following this intriguing trail of indigeneity leads us into the present. The moral and ethical 'cleanliness' of exploration was arguably one of the fundamental but unacknowledged historical conditions that allowed Antarctica to be defined as 'the continent of peace' later in the twentieth century. To this extent, indigeneity can be seen as central, not marginal, to the way we understand and relate to Antarctica in the early twenty-first century.

Notes

1 *Ainu Monogatari* [Ainu Story], a bilingual text in Japanese and Ainu, was published in Tokyo 1913. It is currently being translated into English by Dr Hilary Shibata.
2 Spelled *finnesko* (in both singular and plural) in Swedish, Norwegian and Danish, the anglicised 'Finneskoes' used by some Heroic Era explorers contains a redundant pluralisation that translates as 'Finn's shoes shoes'.
3 Netchelli Eskimos was the name given to the Inuit people of the Netsilik region, in the present-day Canadian Arctic.
4 The guanaco (*Lama guanicoe*) is a camelid, native to Patagonia and Tierra del Fuego.
5 The 'balls and pillars' were reported by Captain Larsen on his Antarctic expeditions in the 1890s. They may well have been geological features called 'concretions' – 'accreted balls of sand and silt and mud cemented with calcium carbonate ... highly rounded rocks ... found in abundance ... along the Antarctic Peninsula' (Stilwell & Long 2011, p. 26).
6 *Terra nullius* is a concept in international law that is often translated as 'unowned land'.

References

Amundsen, R 1912, *The South Pole*, 2 vols, trans. A Chater, Murray, London.
Australian Antarctic Division n.d., 'Australia's new icebreaker – RSV Nuyina', viewed 10 November 2018, <http://www.antarctica.gov.au/icebreaker>.
Borchgrevink, C 1980, *First on the Antarctic continent, being an account of the British Antarctic Expedition, 1898–1900*, 2nd edn, Australian National University Press, Canberra.
Borchgrevink, C 1901, *First on the Antarctic continent, being an account of the British Antarctic Expedition, 1898–1900*, G Newnes Ltd, London.
Cherry-Garrard, A 1965, *The worst journey in the world, Antarctic, 1910–1913*, Chatto & Windus, London.
Cook, F 1980 *Through the first Antarctic night 1898–1899*, 2nd edn, Australian National University Press, Canberra.
Crawford, J 1998, *That first Antarctic winter. The story of the Southern Cross Expedition of 1898–1900 as told in the diaries of Louis Charles Bernacchi*, South Latitude Research Limited in association with Peter J Skellerup, Christchurch.
Dumont d'Urville, JSC 1987, *An account in two volumes of two voyages to the South Seas, to Australia, New Zealand, Oceania 1826–1829 in the corvette 'Astrolabe' and to the Straits of Magellan, Chile, Oceania, South East Asia, Australia, Antarctica, New Zealand, and Torres Strait 1837–1840, in the corvettes 'Astrolabe' and 'Zélée', 1837–1840*, trans. and ed. H Rosenman, Melbourne University Press, Melbourne.
Griffiths, T 2007, *Slicing the silence: voyaging to Antarctica*, New South Press, Sydney.
Gurney, A 1998, *Below the convergence: voyages towards Antarctica 1699–1839*, Penguin Books, London.

Hains, B 2002, *The ice and the inland: Mawson, Flynn, and the myth of the frontier*, Melbourne University Press, Melbourne.

Huntford, R 2009, *Two planks and a passion*, Continuum, New York.

Maddison, B 2014, *Class and colonialism in Antarctic exploration, 1750–1920*, Pickering & Chatto, London.

Maddison, B 2018a, 'People, nature and the Southern Ocean' in F Steel (ed.), *New Zealand and the sea*, Bridget Williams Books, Wellington, pp. 68–88.

Maddison, B 2018b, 'Finnesko' in A Hansen & B Hansen (eds), *Poles apart: fascination, fame and folly*, Royal Society of Tasmania, Hobart, pp. 158–159.

Mawson, D 1996, *The home of the blizzard: an Australian hero's classic tale of Antarctic discovery and adventure*, Wakefield Press, Adelaide.

McCann, J. 2018, *Wild sea. A history of the Southern Ocean*, New South, Sydney.

McKerrow, B 2013, 'Maori first to reach Antarctica?', viewed 22 November 2018, <bobmckerrow.blogspot.com/2013/01/maori-first-to-reach-antarctica.html>.

Newnes, G 1899, 'The Southern Cross Antarctic Expedition', *The Strand Magazine*, vol. 18, no. 105, pp. 278–288.

Nordenskjöld, O & Andersson, J 1977, *Antarctica, or, two years amongst the ice of the South Pole*, 2nd edn, University of Queensland Press, St. Lucia.

O'Hanlon, R 2012, 'Recovering the subject: *subaltern studies* and histories of resistance in colonial South Asia' in V. Chaturvedi (ed.), *Mapping subaltern studies and the postcolonial*, Verso, London & New York, pp. 72–116.

Reynolds, H 1998, *This whispering in our hearts*, Allen & Unwin, Sydney.

Roberts, P 2016, 'The white (supremacist) continent: Antarctica and fantasies of Nazi survival' in P Roberts, L-M van der Watt & A Howkins (eds), *Antarctica and the humanities*, Palgrave Macmillan, London, pp. 105–119.

Russell, L 2012, *Roving mariners: Australian Aboriginal whalers and sealers in the Southern Oceans, 1790–1870*, SUNY Press, New York.

Russell, L 2018, 'Aboriginal Australians as Southern Oceans mariners' in RK Headland (ed.), *Historical Antarctic sealing industry. Proceedings of an international conference in Cambridge 16–21 September 2016*, Scott Polar Research Institute Occasional Publication Series, Cambridge, pp. 73–78.

Samartzis, P 2016, *Antarctica: an absent presence*, Thames & Hudson, Melbourne.

Scott, R 1905, *The voyage of the 'Discovery'*, 2 vols, Macmillan and Co., London.

Scott, R 1907, *The voyage of the 'Discovery'*, 2 vols, Charles Scribner's Sons, New York & Smith, Elder, & Co., London.

Shibata, H 2018, 'Why go so far? The Ainu dog drivers with Lt. Shirase's 1910–1912 Japanese Antarctic Expedition', paper presented at the HASSEG /SCAR conference, Manila, July 2018.

Shirase Antarctic Expedition Supporters' Association (SAESA) 2011, *The Japanese South Polar Expedition, 1910–12: a record of Antarctica*, trans. L Dagnell & H Shibata, Bluntisham Books, Bluntisham.

Spotswood, C 1888, *Voyage of Will Rogers to the South Pole*, Examiner and Tasmanian Office, Launceston.

Stilwell, JD & Long, JA 2011, *Frozen in time: prehistoric life in Antarctica*, CSIRO Publishing, Melbourne.

van der Watt, L-M 2016, 'The whiteness of Antarctica: race and South Africa's Antarctic history' in P Roberts, L-M van der Watt & A Howkins (eds), *Antarctica and the humanities*, Palgrave Macmillan, London, pp. 125–145.

10 Populating Antarctica

Chilean families in the frozen continent

Nelson Llanos

Human settlement in Antarctica has been and remains a challenging task. To date, the partial occupation of the continent has mainly been in the hands of scientists and military personnel, who have lived there temporarily in order to study the complexities of the environment and its ecosystems, or to safeguard the strategic interests of their countries in the region.[1] However, although the harsh natural conditions of Antarctica have hampered the establishment of a large human population, they have not completely prevented the development of ambitious projects for a more extensive occupation of the region.[2] Three decades ago, Chile conducted one of the most notable of these plans. In April 1984, the government of Augusto Pinochet inaugurated Villa Las Estrellas [Town of Stars], a small village located on King George Island. Intended for civilian and military personnel of the Chilean Air Force, this settlement represents a turning point in the history of Antarctica: during their two-year posting, Air Force personnel were accompanied by their wives and children.

Scholarly interest in the origins, development and historical significance of this Antarctic town has been limited. A handful of works that partially address the issue was published in the 1980s by Chilean authors such as Javier Lopetegui (1984, 1986, 1987), a specialist in geopolitics, and Marcia Poupin (1982, 1984, Poupin & Burgos 1994), an architect who promoted Antarctic settlements. This chapter examines the motivations of the Chilean government in carrying out the project of Villa Las Estrellas, and explores the unique experiences of the first families to live in the polar town. It argues that these pioneers helped to create a new social order in Antarctica which, until then, had been built almost exclusively upon men who lived on the frozen continent separated from their loved ones.

The chapter draws primarily on press material from Chile and the United States and from the oral testimonies of some of the first settlers of Villa Las Estrellas, whose accounts were collected through personal interviews with the author between 2016 and 2017.[3] These sources enable a more comprehensive interpretation of the social history of Antarctica, incorporating a bottom-up perspective that draws on the experiences of the Chilean Antarctic settlers, whose names and views have remained outside the historiography until now.

Commercial exploitation of Antarctica?

If settling the icy continent was an extremely complex challenge for the world's great powers, it was even more so for a country like Chile, whose fragile economy and limited technological development have presented a permanent obstacle for much of its history. Nonetheless, since the mid-twentieth century successive governments of the South American nation have assumed the enormous task of populating the so-called *Antártica Chilena* [Chilean Antarctic Territory].[4] Geographical proximity to the continent, as well as the legal and historical rights Chile has claimed over part of it, have long fuelled a desire to consolidate the country's presence in Antarctica (Romero 1985, pp. 5–8).

This goal had to be postponed countless times as a result of inefficient administrative decisions and permanent budget shortages.[5] It was not until the 1940s that definitive steps towards the effective occupation of the Chilean Antarctic Territory were taken. In early 1947, a naval expedition established the first permanent Chilean Antarctic station at Soberanía [Sovereignty], today called Arturo Prat Base (Pinochet 1986, pp. 22–23). Thereafter, through the work of military and scientific personnel, the Chilean government steadily continued its Antarctic activity, with annual expeditions and the construction of new stations. Among these, the most important was Bernardo O'Higgins Base, operated by the Chilean Army. In February 1948, President Gabriel González travelled to Antarctica and personally inaugurated the new station (Villalón & León 2010, p. 213).

Shortly afterwards, O'Higgins Base became the epicentre of Chile's programme on the icy continent, acting as a platform for the development of a more dynamic and modern Antarctic policy, in accordance with the global context of the Cold War (Chilean Army 1948, pp. 9–11). By then, the Chilean Army understood that O'Higgins Base would serve to realise the sovereignty claimed over the territory in Antarctica, and also provide the opportunity to learn how to live on the continent. The central objective was to generate useful knowledge about the Antarctic environment in order to teach future settlers. At that time, it was already projected that the colonisation of the Chilean Antarctic Territory would also be carried out by civilian personnel of the armed forces (such as doctors, nurses, meteorologists and teachers), not only by occupying the territory, but also by exploiting its natural resources (Chilean Army 1949, p. 9).

Before the signing of the Antarctic Treaty, the government of Chile – just like the other countries with interests in the region – conceptualised the frozen continent as an area of economic potential. For that reason, the authorities in Santiago believed that the occupation and colonisation of the Chilean Antarctic Territory could be an effective strategy to contain any foreign attempt to appropriate Antarctic riches. The armed forces' successive expeditions, as well as the establishment of new stations and refuges throughout the 1950s, seemed to indicate that Chilean families would soon be sent to colonise the region (León & Villalón 2018, p. 37). However, the project fell into abeyance following the signing of the Antarctic Treaty in 1959. By suspending territorial claims, this international agreement seemed to discourage the continent's colonisation. As a result, the Chilean project was postponed for more than twenty years.

Unexpectedly, one of the factors that would revitalise Chile's colonisation plan was an economic crisis that started in the Middle East in the early 1970s. In 1973, an embargo on oil exports from Middle Eastern OPEC countries to nations that had supported Israel during the Yom Kippur War led to the 'first oil crisis'. Towards the end of the decade, a new energy crisis emerged as a consequence of the Iranian Islamic Revolution of 1979. Among its clearest effects was the suspension of oil supply to the United States and other Western powers. The global energy crisis thus extended until the end of the decade, demonstrating the profound energy dependence of the United States and its allies on Middle East oil supplies. This situation prompted a search for new hydrocarbon deposits that could satisfy the growing global demand for energy. Soon, Antarctica attracted the attention of the developed world because it was believed that the vast frozen continent could contain massive oil reserves (Gorostegui & Waghorn 2012, p. 212). Scientific studies sustained the idea that oil in areas such as the Ross Sea and the Antarctic Peninsula could put an end to the international energy crisis of the 1970s (*New York Times*, 1 March 1975, p. 39).

At the beginning of the 1980s, such a possibility fuelled the ambitions of the great powers and triggered a wave of international negotiations to initiate the mineral exploitation of Antarctica.[6] This situation put the Chilean authorities on alert and led them to believe that some industrial nations could take the wealth of a territory that the South American country considered, at least in part, to be under its sovereignty.

The project becomes reality

In the hands of a military junta since 1973, the government of Chile understood that it would be very difficult to compete technologically with the great powers in the exploitation of Antarctic resources. Correspondingly, the Pinochet administration began to design a strategy aimed at reaffirming Chile's presence on the icy continent, even without the technical capacity of the industrialised nations. In this way, in late 1970s, Javier Lopetegui, a retired Air Force General and adviser on Antarctic Affairs to the Chilean Air Force (*Fuerza Aérea de Chile* or FACH), started to promote high-level Antarctic tourism. He believed this was the best option to strengthen the country's Antarctic position. By then, and according to Thomas G. Bauer (2011, p. 3), Antarctic tourism was becoming significant due to the 'increasing environmental awareness and consciousness among the people of the developed countries'.

Since the Antarctic Treaty did not prohibit tourism, Chile's proximity to the frozen continent seemed to offer a unique opportunity for the country to increase its Antarctic influence. The epicentre of this project would be King George Island (South Shetland Islands), where FACH had operated Eduardo Frei Base since 1969 (Lopetegui 1984). The Chilean authorities were convinced that, if Antarctica were going to be opened to commercial exploitation, the city of Punta Arenas and

Frei Base should be the gateways (Gorostegui & Waghorn 2012, p. 212). Lopetegui believed that the region could experience something similar to the Alaskan gold rush, and anticipated the creation of a service industry, with hotels, shops, airports and air services for the mining expeditions that would arrive in Antarctica from all over the world (Boudreaux 1984, p. 11). As the international press of the time rightly pointed out, the Chilean project aimed to prepare the country 'for a coming era of mining and tourism while strengthening its claim to a slice of Antarctic territory' (*Chicago Tribune*, 15 November 1984, p. 36).

The first step in that direction was the construction of the modern Lieutenant Marsh Aerodrome, opened in March 1980, within walking distance of Frei Base (Arnello 1987, p. 19). Because of its location, the new aerodrome could connect the Chilean Antarctic Territory with the city of Punta Arenas in just two hours and thirty minutes, and its landing track – which was promptly extended from 900 to 1300 metres – was prepared to accommodate both military and commercial flights. Marsh Aerodrome had snow-removal equipment, bulldozers and off-road transport vehicles, and – as depicted by Javier Lopetegui – was intended to 'compensate the time that Chile had lost in the Antarctic race'. In addition to the new aerodrome, FACH planned the construction of a hotel with a capacity of seventy beds, a dining room, toilets, heating and hot water, which could host tourists, scientists and officials from Chile and the rest of the world (Lopetegui 1984, pp. 164–167).[7] The convenient location of Marsh Aerodrome, as well as its potential as a tourist destination, soon reactivated the historical aspiration to populate the Chilean Antarctic Territory with men, women and children. In this regard, and in addition to the government's plans, intellectuals and academics played a role, arguing that interest in the frozen continent was not only a matter for the authorities, but also a national duty for Chilean citizens.

In 1982, architect Marcia Poupin developed a research project under the auspices of the Urban Studies Institute at the Catholic University of Chile, in which she argued that there were real possibilities to establish permanent settlements in the Chilean Antarctic Territory (Ercilla 1982, p. 7. My translation).[8] In her master thesis, Poupin argued that, as in any other region of the planet, the development of successful human settlements in Antarctica could not be conducted without considering the exploitation of natural resources, because there existed a key 'interdependence relation' between the two. The settlements described by Poupin would be linked to tourism and scientific projects, although without discarding the idea that 'service centres for commercial exploitation of resources' could be established in the future (1982, p. 11. My translation).[9]

That option was particularly important in an international context of growing interest in extractive and commercial activities in Antarctica. Controversially, Poupin asserted that the development of Antarctic settlements, as well as the effective occupation of their surrounding areas, would result in some exercise of territorial and economic sovereignty in geographical spaces that were crucial for accessibility to the frozen continent.[10] As noted by the Chilean architect, this would depend on the location of natural resources, as well as on the level of scientific and technological development of the countries interested in extracting Antarctic riches (Poupin 1984, pp. 115–116).

The studies conducted by Poupin and other scholars, as well as the technical contributions of the Air Force, prompted the Chilean government to design a new National Antarctic Policy, the main goals of which were the construction of new bases and the effective occupation of the country's Antarctic territory. This new strategy, which Gorostegui and Waghorn (2012, pp. 212–214) have described as a 'territorial policy', included the *Proyecto de Asentamiento Poblacional de la Antártica* [Project of Settlement of Chilean Antarctica] to be implemented in 1984.

In collaboration with the Catholic University of Chile and the Housing and Urban Planning Ministry, the Production Development Corporation (Corporación de Fomento de la Producción or CORFO) undertook the design and implementation of the project. FACH, which would assume overall responsibility for this ambitious plan, asserted that its main purpose was to establish a small town that would offer transport and commercial services for international expeditions, and house the future population which would develop to support the activities associated with tourism, science and natural resources exploitation in the Chilean Antarctic Territory (Lopetegui 1984, p. 176).

The Antarctic town, to be located in the vicinity of Frei Base and Marsh Aerodrome, consisted of twenty houses, ranging from eighty to a hundred square metres, which would be built on piles to avoid being buried in the snow. All of them would be equipped with appropriate technology and comforts that would allow the Chilean families to endure the hostile climate while living in Antarctica (*Boletín Antártico Chileno* 1984, p. 62). The project also included a community house with a library and other resources for social activities. Later, the polar settlement would include an infirmary, a school, a gym, a small supermarket, a post office and a bank office (*Morning Star*, 20 December 1983, p. 33).

Antarctic families, Antarctic babies

In order to select the first Chilean families to live on the frozen continent, FACH issued a call to all its members, civilian and military. They were looking especially for married couples who were involved in the professional fields of health, communications and air transport and were able to speak a language other than Spanish.[11] FACH particularly promoted the participation of young couples without children or with children under the age of seven (Camacho 2016, pers. comm., 24 October). As an incentive, and in compliance with the regulations for professional service in isolated regions, FACH offered to pay a salary of five times the usual rate to personnel selected for the mission of living in Antarctica for two years. Couples interested in being part of the project, and who fulfilled the requirements, had to pass a rigorous selection and training process, which included medical and psychological examinations, as well as a gruelling training to learn to ski and survive cold climates. All applicants also took first aid classes and learned techniques for fire control.

After completing the process, FACH selected six families to initiate the Chilean colonisation of Antarctica. The first was the family of Daniel Contreras (an aeroplane pilot) and his wife, Adriana Van Weesel (a medical technologist), and their two children, Andrés and Rodrigo, four and two years old respectively. The second family was that of Germán Camacho (a surgeon) and his wife, Ana María Martino (a nurse), who would travel to Antarctica with their children, Germán (five) and Francisco (four). The third group was the family of Orlando Miranda (a helicopter pilot) and Susana Laguna (a nutritionist), parents of Gabriela, only eight months old. Adolfo Cortés (a meteorologist) and Gisella Rojas (a member of the Aeronautics Division), a young couple recently married, without children, joined the group in fourth place. An air traffic controller, Reineri Merino, and his wife, Nilda Álvarez (a housewife), were the fifth family participating in the project. They would travel to Antarctica without their children, who were already at high school. The last family selected was Domingo Rojas (an air traffic controller), his wife, María Teresa Zúñiga (a nutritionist), and Gonzalo, their seven-year-old son. In total, the group of pioneers consisted of eighteen people: twelve adults and six children (*El Mercurio de Valparaíso*, 26 January 1984, p. 10).

The six families took off from Santiago Airport in the direction of Punta Arenas on 14 February 1984. Of their arrival in Punta Arenas, the local newspaper reported that all the Antarctic settlers 'looked happy', as they waited anxiously and expectantly for the next leg of the journey that would take them to their 'home and habitat' for the next two years (*La Prensa Austral*, 15 February 1984, p. 9. My translation).[12] The plan had been to travel to the frozen continent on the same day; however, the weather conditions, particularly strong winds, prevented the C-130 Hercules from landing at the Marsh Aerodrome, and the families were forced to return to Punta Arenas (*El Mercurio de Valparaíso*, 15 February 1984, p. 9). Eventually, the settlers would arrive in King George Island on 15 February at 3pm – the exact moment at which their historic experience on the frozen continent began.

The official inauguration of the Chilean Antarctic town – Villa Las Estrellas – took place on 9 April 1984. The President of the Republic, General Augusto Pinochet, attended the ceremony accompanied by his wife and a large retinue, including (among others) the commander-in-chief of FACH; the respective Ministers of National Defence, Foreign Affairs and Public Works and Housing; the Secretary General of the Government; and the Vice President of CORFO (*El Mercurio de Valparaíso*, 10 April 1984, p. 8). In a speech to the world's media Pinochet declared:

> Today, men, women and children of Chile live in this village … culminating an effort of decades … You will have to face the solitude and the inclemency of a harsh environment, while you work, study and make a life like the other Chilean families. You will prove with your presence that human life in Antarctica is possible and can be enriching, [opening] wide roads and wide horizons for our country and the whole world.
>
> (*Boletín Antártico Chileno* 1984, p. 60. My translation)[13]

By arguing that Antarctica was 'the reserve of the world', which offered 'endless promises' for humankind,[14] Pinochet made clear that the purpose of Villa Las Estrellas was directly connected to the potential exploitation of natural resources in Antarctica (*El Mercurio de Valparaíso*, 10 April 1984, p. 8. My translation).

Captain Daniel Contreras – commander of Frei Base and mayor of the polar town – also expressed his feelings during the ceremony. The leader of the settlers declared to the national and international press that this project was 'of great importance for the formation of future population groups' in Antarctica, and would encourage the development of other activities in the region, like those linked to the exploitation of natural resources (*Boletín Antártico Chileno* 1984, p. 61. My translation).[15]

The impact of the Chilean initiative would be even greater when FACH announced the arrival of three new settlers. According to the testimonies of the founders of Villa Las Estrellas, Ana María Martino passed out during the inauguration ceremony – the first sign of her pregnancy. Before long, two other women – Gisella Rojas and Susana Laguna – would make their own contribution to the population boom in the polar town. The news of the Antarctic pregnancies was kept from the press for a long time, and was only reported when the date of the first birth approached. By mid-November, the *Chicago Tribune* informed the American public of the birth of the first three 'native Antarcticans' and the arrival of seven new families for the coming year. The newspaper highlighted Chile's audacious Antarctic project and criticised the limitations of US actions in the icy continent, observing that

> While the United States treats Antarctica as a pristine laboratory for scientific research, and other nations, like Japan, rush to fish and look for oil off its shores, only Chile and Argentina are populating it with families.
> (*Chicago Tribune*, 15 November 1984, p. 36)

Juan Pablo, the first Chilean born in Antarctica, came into the world on 21 November 1984. A complete medical team travelled from Punta Arenas specially to monitor the delivery. According to Dr Camacho, father of the 'little penguin'[16] – as his family nicknamed baby Juan Pablo – members of the nearby bases, among them some Argentines, Germans, Poles and Soviets, were very happy about the new settler in the Chilean Antarctic town (2017, pers. comm., 24 October). In an official statement, FACH announced that, 'We are pleased to report to the public the first birth of a Chilean citizen in Villa Las Estrellas … mother and baby are doing well (*La Prensa Austral*, 22 November 1984, p. 22. My translation).[17]

Days later, on 2 December 1984, the polar town welcomed baby Gisella, daughter of the Cortés-Rojas family, and the only Chilean girl born in Antarctica. Finally, on 23 January 1985, Ignacio – son of Orlando Miranda and Susana Laguna – was born. Although pregnancies would continue to occur in Villa Las Estrellas, the government decided that, for safety and financial reasons, all new births should take place in the city of Punta Arenas (Van Weesel 2017, pers. comm., 8 April). That decision was the first sign that the Chilean settlement was not the beginning of an effective colonisation process of the frozen continent; rather, it was a symbol of the country's sovereignty aspirations in Antarctica.

Living on the icy continent

The settlers of Villa Las Estrellas faced an unknown and challenging reality, living through unique experiences while they were part of the Chilean Antarctic project. By delving into their memories, more than thirty years after their departure from the frozen continent, it is possible to identify that the primary difficulties they had to overcome were linked to geographical isolation; to a constant feeling of danger; and, of course, to their daily interaction with other families and authorities, in a small society under construction, where the norms of coexistence and the distribution of power were not always agreed or accepted by all settlers.

First, and although FACH publicly maintained that isolation 'was a thing of the past', the Chilean authorities knew that such a problem, as well as the monotonous daily routine, could generate disturbing levels of uncertainty and anxiety, and even mental disorders, among the Antarctic pioneers (Handler 1983, p. 13). A team of psychologists had examined the settlers before they started their adventure, to determine whether they would be able to withstand the challenges of life on the frozen continent. During their training, the settlers learned techniques to 'kill time' while living in polar solitude through handicrafts, group games and other activities (Van Weesel 2017, pers. comm., 8 April).

Although the families of Villa Las Estrellas were certain that the experience would be demanding, they did not want to fail. Nilda Álvarez, one of the women in the group, explained that all the settlers were aware of what the isolation implied, but that 'they were prepared and trusted in God' so that everything 'would come out well' (*El Mercurio de Valparaíso*, 15 February 1984, p. 9. My translation).[18] In statements to the Chilean press, Orlando Miranda confirmed that his family, and the rest of the settlers, had spent several months in preparation courses, receiving psychological support to face the Antarctic isolation, as well as the personal and marital problems that could arise (*La Prensa Austral*, 15 February 1984, p. 9. My translation).[19]

With some success, the pioneers also designed their own strategies to face their daily routine and keep up morale. They organised football championships; formed a folk music group; ran discussion meetings on topics like history, culture and current affairs; and implemented a news channel that was transmitted on closed-circuit television to all the houses. They even created a hand puppet who advised the children to 'maintain good behaviour, brush their teeth and do their homework'[20] – every night, the children of Villa Las Estrellas looked forward to the puppet appearing on their TVs to wish them good night (Van Weesel 2017, pers. comm., 8 April).

Despite the families' best efforts, the Antarctic isolation, the remoteness and the daily routine seemed ever-present. Information and news about Chile and the world arrived with a delay of weeks or months; the settlers could only communicate with their families and friends once a week; and they had to accept that for two years they would be forced to live with the same people and share the same areas. Of particular concern was food. In addition to learning how to treat and ration water, the Chilean settlers had to learn to cook with frozen and canned food, something that was not common in Chile at that time. Provisions were supposed to arrive once a month from Punta Arenas, but things did not always go according to plan. Several of the settlers remember that the weather sometimes delayed the arrival of food, and that during a harsh winter the aeroplane could not land even once at Marsh Aerodrome, generating a shortage of food and profound uncertainty among the families. Recalling his childhood in Villa Las Estrellas, Gonzalo Rojas states that, when food was scarce, men went fishing in the vicinity of the town: 'We ate *nototheniidas* … they did not taste good, but it was the only thing we had to eat' (Rojas 2016, pers. comm., 19 April. My translation).[21] On one of those fishing trips, Gonzalo remembers that 'the group was surprised by a terrible storm, which almost prevented them from returning home, leaving them semi-buried in the snow' (Rojas 2016, pers. comm., 19 April. My translation).[22]

In fact, along with the problems generated by isolation, the harsh Antarctic environment turned out to be one of the most serious difficulties the settlers had to face. Although Villa Las Estrellas was situated in a relatively safe location (compared with the rest of the continent), every day the environment reminded the settlers that they were in a challenging and dangerous place, where death was a constant threat. Gonzalo, the eldest of the village children, remembers that for all of them it was a very hard experience because they lived 'with the constant fear of getting lost or falling into a crack' (2016, pers. comm., 19 April. My translation).[23] He recalls a near-fatal accident that his father, Domingo Rojas, suffered on the landing track of Marsh Aerodrome, when a steel cable hit his head. As soon as they heard what had happened, Gonzalo and his mother rushed to the site of the accident and saw that the snow around Domingo had turned deep red with his blood. The injured man spent months in bed, unable even to feed himself. The damage to one of his ears would be permanent, forcing Domingo to abandon his career soon after (Rojas 2016, pers. comm., 19 April).

Germán Camacho, who provided first aid to Domingo Rojas after his accident, recalls that one of his worst experiences as a doctor also happened on the frozen continent. In the first days of 1986, with only a few weeks left before he returned to his 'normal life' in Santiago, there was a terrible plane crash near Villa Las Estrellas. It was a private aeroplane carrying eight US tourists to celebrate the New Year in Antarctica. They had planned to land at Marsh Aerodrome, but low cloud and bad weather prevented it. Camacho and the other settlers remember that they could hear the engine of the plane when it tried to land, and that they could even see its fuselage as it approached the landing strip (Camacho 2016, pers. comm., 24 October). Although the plane also tried to land at Marambio Station (operated by Argentina), it was soon confirmed that it had crashed on Nelson Island, not far from the Chilean Antarctic town. Dr Camacho, along with the pilot Orlando Miranda and other members of FACH, found the wreckage and recovered the bodies of the passengers and the crew – ten people in total – which were taken to Frei Base (James 1986, p. 1). Gonzalo Rojas remembers that, disobeying his parents, he went to the place and saw how the bodies were buried in the snow, to preserve them until they could be sent to Punta Arenas (2016, pers. comm., 19 April).

A third challenge defined the Chilean settlers' stay on the icy continent: trying to build a 'normal society' in a unique and hostile environment that was dominated by men and where tension between civilians and military was permanent. In this small town, which replicated traditional Chilean society of the time, established gender roles were strictly defined. Thus, while the men went to work in the morning at Frei Base, the women were engaged in housekeeping and caring for the children.[24] The women of the village also had to take charge of the post office, the library and the tourist shop, and – unexpectedly –[25] teach their own children. While the older children were taught subjects like history, biology and maths, the village women were also responsible for teaching the basics skills of reading and writing to the youngest pupils. Adriana Van Weesel remembers that the Minister of Education provided manuals and teaching materials as a makeshift solution until the school officially opened the following year, with two professional teachers (2017, pers. comm., 8 April).

During the first year of Villa Las Estrellas, the excessive workload of the women added more tension to the differences between the civilian and military personnel. Although the highest authority of the town was a military captain (Contreras), most of the settlers were civilians (including all the women). They were not used to following military rules, let alone implementing those rules within their homes. In short, the settlers felt that there was no clear separation between the public and private spheres. According to an article published in the *New York Times*, the excessive military discipline imposed in the polar town only ended when the women organised themselves and demanded appropriate treatment for the civilian settlers. In particular, the 'housewives' obtained more time to devote to their families by making their working time in the school and the other units more flexible (*New York Times*, 4 December 1984, p. 17). However, unlike their husbands – and although most of these women were professionals and civilian members of FACH – they did not receive a salary while they were living in Antarctica.

Unexpected friendships

Looking back on their days in Antarctica, the founders of Villa Las Estrellas assert that, although coexistence among the settlers was not always harmonious, they did have cordial and friendly relations with members of the nearby bases. At that moment, the political context was extremely complex: under pressure from the United States to put an end to the dictatorship, the right-wing military junta headed by Pinochet was diplomatically isolated. Chile was also maintaining bellicose relations with Peru, Bolivia and especially Argentina (Mellafe 2018). At the same time, Chile remained one of the last Latin American bastions against international communism in the final years of the Cold War. Ironically, the settlers sent to Antarctica by Pinochet's authoritarian government built strong and fraternal bonds with those who were supposed to be their enemies.

In November 1984, Associated Press journalist Richard Boudreaux visited the Chilean polar town. Expressing his surprise at the international cooperation he observed in the daily life of the Antarctic community, Boudreaux noted that, 'The most striking sign of this cooperation was the bear hug exchanged at frequent social gatherings between the Chileans and their Soviet neighbours at the Bellingshausen scientific station' (1984, p. 11). Effectively, the settlers of Villa Las Estrellas built a peaceful and cordial coexistence, with both the Soviets and the East Germans who worked at the Soviet station. According to Boudreaux's report, Captain Contreras himself had pointed out that Chileans and Soviets 'could not afford to bring their political differences to the harsh Antarctic environment' (p. 17).

Three decades after those statements, Contreras reaffirmed, at his house in Santiago, that his family maintained an excellent relationship with the Soviets; they shared and exchanged the products they needed, celebrated holy days together and learned from their cultural differences. They also formed a strong friendship with some scientists from the German Democratic Republic (GDR): 'they visited us in Chile after the fall of the Berlin Wall. Then we visited them in Germany' (Contreras 2016, pers. comm., 8 April. My translation).[26] Confirming this account, Martin Kaiser, one of the closest German friends of the Contreras family – who visited them again in Chile in early 2017 – asserts that scientists at Bellingshausen Station maintained warm relations with the Chileans in spite of the political context. The biologist remembers that, since Chile and the GDR did not maintain official relations, he and the other scientists had to be very cautious about their closeness to the South Americans. A political officer at Bellingshausen Station kept this singular friendship under constant surveillance (Kaiser 2017, pers. comm., 27 February).

From a similar perspective, Camacho remembers that his son Juan Pablo, born in Antarctica, fell asleep in the arms of the Soviets and Germans almost every day. The doctor of Villa Las Estrellas confirms that friendship, fraternity and cooperation were common in Antarctica. At his house in Villa Las Estrellas, he received Polish, Soviet and even Cuban visitors at a time when the Washington–Moscow conflict still kept the world divided. 'Once I broke my foot and I got treatment at the Soviet station', he said. 'Can you imagine it? The injured doctor of the Antarctic village of Pinochet, treated by a communist surgeon. In Antarctica the logic of the Cold War did not exist' (Camacho 2017, pers. comm., 24 October. My translation).[27]

After two years living on the most hostile continent on earth, the founders of Villa Las Estrellas left the small town on 23 January 1986. By then, several new families had already settled in the village. Although the Antarctic town continued to operate for more than thirty years, in June 2018 the Chilean government announced its temporary closure in order to rebuild the houses and solve other existing infrastructure problems. Juan Pablo, Gisella and Ignacio have never returned to the land of their birth.

Conclusions

The town of Villa Las Estrellas is the realisation of one of the fundamental goals of Chile in Antarctica – the effective occupation and colonisation of its territorial claim. In this sense, it is the most important initiative in the country's Antarctic history since the 1940s. The project reactivated Chile's Antarctic policy at a particularly complex moment, in which the authorities and the armed forces perceived that the Antarctic Treaty System – and consequently the country's rights over part of the continent – were threatened by the potential exploitation of Antarctic resources.[28] By developing tourism services and sending families to occupy the so-called *Antártica Chilena*, the government tried to strengthen Chile's presence in a region that the South American country has long considered part of its territorial heritage.

As the Antarctic Treaty's signatory countries ultimately prohibited commercial exploitation of the continent's riches, the historical significance of Villa Las Estrellas can now be understood in a different dimension. Although for decades scientists and military have explored, worked and lived in Antarctica, the presence of families on the continent has been highly unusual. The singular experience of the Chilean settlers of Villa Las Estrellas constitutes an extraordinary case of families trying to develop a relatively normal life in the middle of the Antarctic wilderness, including giving birth and raising children. Only Argentina has attempted a similar project in Fortín Sargento Cabral, near the Antarctic Peninsula.[29]

The arrival of the Chilean families immediately altered (but did not replace) the social order that had existed in Antarctica since its discovery: a region inhabited mainly by men, with very few women and without children. Along with the settlers living in the Chilean town, new activities and institutions suddenly appeared in the Antarctic landscape, all of which were required for sustaining this fledgling society. Villa Las Estrellas laid the groundwork for the development of a more traditional structure of social relations in Antarctica, just as Marcia Poupin and other scholars had projected. The Chilean polar town proved that, despite the evident geographical singularities, it was possible to replicate community life in Antarctica, just as it is in other small towns in climatically hostile regions of Chile and the world.

In summary, the audacious Chilean project demonstrated for more than three decades that it is possible to develop family life in Antarctica, in a way that is in harmony with citizens of other countries and that overcomes the obstacles of the natural environment. Such a valuable experience, which, of course, included the virtues and vices of every human community, can also serve as a model for future settlement processes in the region. At the same time, new studies are needed to expand our knowledge about the social aspects of human presence on the icy continent, covering topics such as the history of the family, children and women, and all peoples who have remained outside the history of Antarctica so far.

Notes

1 Although the Antarctic Treaty prohibits military bases and manoeuvres, it allows the use of military personnel or equipment on the continent for scientific research or for any other peaceful purpose.

2 Unlike other cases, Chile has never officially laid a territorial claim in Antarctica. From the South American country's perspective, the Chilean Antarctic Territory has been part of the national territory since the origins of the republic, as a result of historical, legal, and geographical reasons. Therefore, the project of populating Antarctica was just one more step in the long process of consolidating Chile's territorial sovereignty, which – over time – also included the northern frontier, Patagonia, and the Pacific Ocean, among other regions.

3 All translations from Spanish language articles and interviews are the author's.

4 The Chilean Antarctic Territory (a province of the country and not a territorial claim, from Chile's perspective) covers a total of about 1,250,000 km^2, and ranges from 53° west to 90° west, as far as the South Pole. The Chilean government argues that this territory was inherited from the Spanish empire, which – according to the Treaty of Tordesillas (1494) – possessed all lands to the west of a line running along the Atlantic Ocean, from pole to pole. The Chilean Antarctic Territory partially overlaps the British and Argentine claims in the Antarctic Peninsula area.

5 Chile's first official attempts to explore and occupy the Antarctic continent took place in 1906, especially due to the influence of Antonio Huneeus Gana, Secretary of Foreign Affairs at that time.

6 Some serious discussions about exploitation of Antarctic minerals had already taken place during the Ninth Antarctic Treaty Consultative Meeting in London (1977). Later, the Convention on the Regulation of Antarctic Mineral Resource Activities (CRAMRA) would seek to manage extractive activities in Antarctica. However, signatory states did not ratify it, instead replacing it with the Protocol on Environmental Protection to the Antarctic Treaty (Madrid, 1991).

7 Frei Base and its aerodrome remain the most important tourist gateway to Antarctica. The small hotel continues to operate.

8 'Hay posibilidades reales para los asentamientos que – hipotéticamente – se instalen en el continente helado … Se proyecta una ciudad que, en la próxima década, será el centro de la estrategia de desarrollo urbano regional.'

9 'La actividad económica constituye la única manera por la cual una región como ésta … puede ser habitada estable y masivamente por el hombre … El objetivo es dar los lineamientos para desarrollar asentamientos humanos antárticos capaces de sostenerse económicamente con la explotación de los recursos naturales existentes, en el entendido que los asentamientos y la explotación son interdependientes.'

10 This was controversial because the Antarctic Treaty established that no claim could legally be reinforced when the Treaty was in force. From Chile's perspective, however, its Antarctic territory is not a claim, but a province of the country. As a result, the government argued that the Antarctic settlement was going to be developed in area under its sovereignty.

11 Several international stations were located in the vicinity of King George Island, so speaking more than one language was vital for settlers of Villa Las Estrellas, especially in an emergency.

12 'A su llegada a Punta Arenas, todos los integrantes de este grupo de familias, se mostraron felices, ansiosos, y expectantes, ya en el tramo final de su viaje hasta lo que será su hogar y hábitat durante dos años.'

13 'Hoy, hombres, mujeres y niños de Chile viven en esta villa … culminando así un esfuerzo de decenios … Deberán enfrentar ustedes la soledad y las inclemencias de un medio difícil, mientras trabajan, estudian y hacen una vida como las demás familias chilenas … Probarán con su presencia que la vida humana en la Antártica es

posible y puede ser enriquecedora, abrirá anchos caminos y amplios horizontes para nuestro país y el mundo entero.'

14 'Esta zona es vista como la reserva del mundo … y encierra horizontes y promesas inagotables para nuestros hijos.'

15 'Las proyecciones del plan de colonización en la zona revisten una significativa importancia para la formación de futuros núcleos poblacionales, lo que contribuirá a fomentar el desarrollo paulatino de actividades de diferente índole … En este territorio se encuentran insospechadas reservas económicas para el future de Chile y el mundo.'

16 'Pingüinito.'

17 'Tenemos el agrado de informar a la opinion pública el feliz nacimiento del primer ciudadano chileno en la Villa Las Estrellas … La madre y su hijo se encuentran en perfectas condiciones.'

18 'Estamos perfectamente conscientes de lo que significa a partir de este momento al aislamiento. Estamos preparados. Que Dios nos acompañe.'

19 'Respecto a posibles problemas de parejas … señaló que habían tenido cursos de sicología y sociología, así que vamos con terapias personales para subsanar … ese tipo de problemas.'

20 'Hicimos un títere que les decía a los niños que se portaran bien, que cepillaran sus dientes, y que hicieran sus tareas.'

21 'Comíamos nototenias … no tenían buen sabor, pero era lo único que había para comer.' *Notothenias* and the rest of the notothenioid fishes are abundant in Antarctic waters. As noted by Amores et al., they are known for having evolved key genetic innovations in cold waters, including 'the origin of antifreeze glycoproteins and a greatly altered inducible heat shock response' (2017, p. 2196).

22 'El grupo fue sorprendido por una tormenta terrible que casi nos impidió el regreso a casa, y nos dejó semi-sepultados en la nieve.'

23 'Vivíamos con el miedo constante de perdernos o caer en una grieta.'

24 The government had told the women that they would continue with their professional roles in the Air Force in Antarctica; however, due to budgetary constraints, it was decided that they would not be paid.

25 The project originally included professional teachers for the school in Villa Las Estrellas during its first year. Again due to budgetary issues, they only started work in 1985. The settler families were not informed of this change until a few weeks before their arrival in Antarctica.

26 'Ellos nos visitaron en Chile luego de la caída del muro de Berlín. Después, nosotros los visitamos en Alemania.'

27 'Una vez me quebré un pie y recibí tratamiento en la base soviética … ¿Puede imaginarlo? El médico de la ciudad antártica de Pinochet atendido por un cirujano comunista. En la Antártica la lógica de la Guerra Fría no existía.'

28 Traditionally, Chilean authorities (as well as those of Argentina) have interpreted the Antarctic Treaty – especially Article IV – as a legal tool that safeguards the country's rights in the continent.

29 The Argentine government inaugurated Fortín Sargento Cabral in February 1978. The small Antarctic settlement was built for military personnel, their wives and children. The first families living there included several women in the last months of their pregnancies.

References

Boletín Antártico Chileno 1984a, 'Chile Consolida su Soberanía', vol. 4, no. 1, pp. 59–62.

Chicago Tribune 1984, 'Just another day in Antarctica. Colonists make chilly Chilean outpost a home', 15 November, p. 36.

El Mercurio de Valparaíso 1984, 'Colonizadores antárticos siguen en Punta Arenas', 15 February, p. 9.

El Mercurio de Valparaíso 1984, 'S.E. inauguró la primera villa antártica chilena', 10 April, p. 8.

El Mercurio de Valparaíso 1984, 'S.E. viaja a la Antártida', 26 January, pp. 1, 10.

Ercilla 1982, 'Una idea para el futuro', no. 2443, p. 7.

La Prensa Austral 1984, 'Ansiosos por llegar los nuevos colonos', 15 February, p. 9.

La Prensa Austral 1984, 'Inauguración Villa Las Estrellas', 12 April, p. 4.

La Prensa Austral 1984, 'Juan Pablo se llamará el primer chileno antártico', 22 November, p. 22.

Morning Star 1983, 'Frozen colony. Chilean pioneers going to Antarctica', 20 December, p. 33.

New York Times 1975, 'Big Antarctic oil field is possible, Navy says', 1 March, p. 39.

New York Times 1984, 'Chilean families begin colony on Antarctica', 4 December, p. 17.

Amores, A, Wilson, CA, Allard, CHA, Detrich, HW & Postlethwait, JH 2017, 'Cold fusion: massive karyotype evolution in the Antarctic Bullhead Notothen Notothenia coriiceps', *G3: Genes, Genomes, Genetics*, vol. 7, no. 7, pp. 2195–2207.

Arnello, M 1987, 'Perspectivas futuras del territorio antártico chileno', *Revista Chilena de Geopolítica*, vol. 3, no. 2, pp. 11–20.

Bauer, T 2011, *Tourism in the Antarctic. Opportunities, constraints, and future prospects*, Routledge, New York and London.

Boudreaux, R 1984, 'New pioneers test family life in Antarctica', *Schenectady Gazette*, 21 November, p. 11.

Chilean Army 1948, *Base O'Higgins. Territorio Antártico Chileno*, Instituto Geográfico Militar, Santiago.

Chilean Army 1949, 'Vida de la primera guarnición militar antártica', *El Memorial del Ejército de Chile*, vol. 43, no. 230, pp. 9–19.

Gorostegui, J & Waghorn, R 2012, *Chile en la Antártica. Nuevos desafíos y perspectivas*, USACH Ed., Santiago.

Handler, B 1983, 'Chile to colonize Antarctica with families', *Gainesville Sun*, 26 December, p. 13.

James, G 1986, '8 U.S. tourists killed as plane crashes on an island in Antarctic', *New York Times*, 2 January, p. 1.

León, C & Villalón, E 2018, *Chilenos en la Antártica. Base O'Higgins, 1948–1958*, Chilean Army Editions, Santiago.

Lopetegui, J 1984, 'Infraestructura antártica y política de acceso al continente' in F Orrego, M Infante & P Armanet (eds), *Política Antártica de Chile*, Instituto de Estudios Internacionales, Universidad de Chile, Santiago, pp. 161–177.

Lopetegui, J 1986, *Antártica, un desafío perentorio*, Ediciones Génesis, Santiago.

Lopetegui, J 1987, 'Finalidad, objetivos y metas de la utilización antártica', *Revista Chilena de Geopolítica*, vol. 4, no. 1, pp. 158–164.

Mellafe, R 2018, *Al borde de la guerra. Chile-Argentina 1978*, Legatum Editores, Santiago.

Pinochet, O 1986, *Base Soberanía y otros recuerdos antárticos*, Editorial Andrés Bello, Santiago.

Poupin, M 1982, 'Asentamientos antárticos y explotación de recursos: Una relación de mutua dependencia', unpublished Masters thesis, Catholic University of Chile, Santiago.

Poupin, M 1984, 'Poblamiento Antártico' in F Orrego, M Infante, & P Armanet (eds), *Política Antártica de Chile*, Instituto de Estudios Internacionales, Universidad de Chile, Santiago, pp. 115–118.

Poupin, M & Burgos, L 1994, 'Asentamientos antárticos, un desafío para la relación sociedad-naturaleza', *Boletín Antártico Chileno*, vol. 40, no. 4, pp. 2–7.

Romero, P 1985, *Síntesis de la historia antártica de Chile*, Colección Terra Nostra, Universidad de Santiago, Santiago.

Villalón, E & León, C 2010, *Jalonando Chile austral antártico*, Chilean Army Editions, Santiago.

11 Placing the past

The McMurdo Dry Valleys and the problem of geographical specificity in Antarctic history

Adrian Howkins

Introduction

In an essay urging humanities scholars to think more carefully about place in Antarctica, historian Alessandro Antonello (2016, pp. 181–204) notes that researchers in these disciplines tend to apply a sense of 'placelessness' to their studies of the continent. There are few other parts of the world, he convincingly argues, that are treated quite as generically as Antarctica. This flattening of place is potentially exacerbated by thinking about Antarctica through the lens of the Anthropocene. Although it is certainly useful as a way of acknowledging the extent of human impact on the natural world, the Anthropocene is, by definition, universal and universalising (Lecain 2015). Its very existence as a concept is based on the proposition that humanity (or at least a section of humanity) has fundamentally altered the whole of the Earth, leaving little scope for local differences or exceptions.

This chapter uses the McMurdo Dry Valleys near Ross Island to consider the problem of geographical specificity in Antarctic history. The McMurdo Dry Valleys are the largest predominantly ice-free region in the Antarctic continent. Whereas much of the rest of the continent is covered in a thick ice-sheet, the Dry Valley region consists of bare rocks and soils interspersed with glaciers and ice-covered lakes. For a few weeks during the twenty-four-hour sunlight of summer, meltwater streams flow from the alpine glaciers that line the valleys' walls. Liquid water supports the existence of a variety of often-microscopic ecosystems in the lakes, streams, soils and even in meltwater cryoconite holes on the glaciers themselves. The Dry Valleys are dry as a result of their location in the Transantarctic Mountains. In this part of the continent, the mountains block the advance of the East Antarctic Ice Sheet at the same time as channelling the powerful katabatic winds that keep the region largely free from snow accumulation. Although precipitation does occur, especially in the coastal regions, the McMurdo Dry Valleys are a classic cold desert environment (Priscu 1998).

As a result of these stark environmental differences from the surrounding region, the McMurdo Dry Valleys offer a useful location for thinking about the construction of place in Antarctic history. Differences, similarities and connections stand out here in ways that are often harder to see in regions where environments are more alike. This chapter builds on existing scholarship on place to ask two broad questions: how can we write engaging, relevant histories of Antarctica in places such as the McMurdo Dry Valleys without falling into the trap of flattening the rest of the continent? And how should we respond to the idea of the Anthropocene in the history of Antarctica, especially as it relates to the idea of place? In order to address these questions, this chapter examines three inter-connected themes from the history of the McMurdo Dry Valleys – political claims, scientific research and environmental management – to illuminate how ideas of place have been constructed. This approach highlights the fact that, in thinking about the meaning of the McMurdo Dry Valleys, place has been an integral and frequently contested part of the continent's wider history.

The McMurdo Dry Valleys region was first sighted by Britain's Captain Robert Falcon Scott in December 1903, at the very end of the *Discovery* expedition. He famously described what he saw as a 'Valley of the Dead', noting that he saw no living thing during the few hours he spent there (Scott 2001, p. 567). The Dry Valley was further explored by a team of scientists led by Australian geologist and geographer Griffith Taylor during Scott's *Terra Nova* expedition in February 1911 (Howkins 2016). Following Taylor's expedition, there was a break of over forty years before the next scientific parties went into the McMurdo Dry Valleys region as part of the International Geophysical Year (IGY) of 1957–58 (Bull & Barwick 2009). Since this renewal of interest, scientists have worked in the region every summer season (and over the course of three winters) since the late 1950s, making the Dry Valleys one of the most important locations for scientific research in Antarctica. Among the many scientists attracted to the region, geologists are drawn by its bare rocks and soils, ecologists by the presence of 'end member' ecosystems existing in an extreme environment and glaciologists by the existence of moraines, erratics and other evidence of past glaciations (Priscu 1998).

The McMurdo Dry Valleys are clearly a very different environment from the surrounding region; yet, the scientists working there have frequently made wider claims for the implications of their research. Geologists have used studies in the McMurdo Dry Valleys to learn about the geological history of Antarctica more broadly; ecologists have used the relative lack of complexity to formulate ecological theory and to consider ecosystem response to a changing climate; glaciologists have gathered evidence to help reconstruct past glaciations (Priscu 1998). On one level, this approach to using scientific research in the McMurdo Dry Valleys to address broader questions makes a lot of sense. In a continent largely covered by thick ice-sheets, the region certainly offers a rare opportunity to see below the ice. It is also relatively accessible from the Antarctic logistics hub of the United States' McMurdo Station and New Zealand's Scott Base on Ross Island. On another level, however, there is something quite paradoxical about using a region which differs from much of the rest of Antarctica to address questions about the continent as a whole. While there might be an appreciation for local specificity and com-plexity in the study area itself, such nuance cannot easily be applied to wider stu-dies, which assume a certain homogeneity across the entire continent.

The strategy of using scientific research in one location to address broader questions is not, of course, unique to the McMurdo Dry Valleys. In a sense, it is part of the fundamental nature of scientific research, especially in the environmental sciences (Bowler 1993). But there does seem to be something particular about the polar regions that encourages broad claims to be made for the relevance of the research being conducted. Writing about the Canadian Arctic, historian of science Steven Bocking (2007) has argued that Arctic scientists have become adept at making claims for the wider importance of their research. Research in the polar regions often lacks the obvious value to humanity of research conducted closer to large population centres, and scientists frequently find themselves making broader claims to justify the elevated expense of working in the Arctic and Antarctica. This tendency is highlighted by the current tagline of the British Antarctic Survey (2015): 'Polar Science for Planet Earth'. Scientists continue to play a powerful role in the history of Antarctica: if they frequently make assertions that flatten place, it is hardly surprising that others do, too.

As a result of its universalising tendency, the concept of the Anthropocene has the potential to exacerbate the trend towards flattening place in Antarctica. But a careful examination of this idea in the context of the history of a place like the McMurdo Dry Valleys immediately complicates the story. In the McMurdo Dry Valleys – depending on what date, if any, is chosen for the beginning of this proposed new geological epoch – it is possible that humans were fundamentally altering the environment before anyone ever set foot in the valleys (Carey et al. 2014). The idea that humans do not even need to be present in a region to have an impact says much about the capacity for human activities to be an agent for global change. At the same time, however, despite its potential utility, the term 'Anthropocene' itself is largely conspicuous by its absence in even the most recent scientific literature on the region, suggesting perhaps that such a broad-brush concept is of little value in the actual work of understanding the environment.

By revealing both the complexity and specificity of place, the history of the McMurdo Dry Valleys would seem to suggest that every region in Antarctica should be examined on its own terms. Even in this relatively small region, multiple overlapping boundaries have frequently existed, and various factors have come together to construct different ideas of place. The history of the McMurdo Dry Valleys shows that broader trends have shaped this region, at the same time as events from this region have shaped wider histories. So, just as it is important to think about specificity, it is also important not to look at any place in isolation: global and continental interconnections are very much part of regional and local histories. These interconnections open the way for thinking about the Anthropocene in relation to the construction of a region like the McMurdo Dry Valleys. This idea, however, needs to be included in a way that takes into consideration local specificity. A one-size-fits-all vision of the Anthropocene does not seem appropriate for thinking about the past, present or future of a continent where we are only just beginning to appreciate the richness and diversity of place.

The politics of place

Although less bitterly disputed than the Antarctic Peninsula region, where Britain, Argentina and Chile make conflicting sovereignty claims (Fontana 2014; Howkins 2017a), the McMurdo Dry Valleys have a contested political history. Since 1923, the region has been claimed by New Zealand as part of the Ross Dependency (Templeton 2000). This claim, however, is not recognised by the United States or Japan, the two other countries with the most significant historical interests in the region. In this contest over sovereignty, the meaning of the McMurdo Dry Valleys as a place has played an important role. New Zealand and the United States – as well as Japan to a lesser extent – have sought to define the McMurdo Dry Valleys in ways that suit their respective political interests. Geographer Klaus Dodds (2002) has discussed how the naming of Antarctic geographical features has exerted a particularly strong political significance in the Antarctic Peninsula region; this is also the case in the McMurdo Dry Valleys.

As in other parts of the Antarctic continent, the early Heroic Era history of the region has had a disproportionate influence on the subsequent political history of the McMurdo Dry Valleys and on the construction of place. The expeditions of Scott and Taylor not only 'discovered' the region, but also gave names to many of the features, which frequently remain in use today. Much of the early politics of naming was intensely personal. Taylor, for example, went to great lengths, in an extended dispute with Hartley Ferrar, another Heroic Era geologist (Debenham 1923), to ensure the glacier and valley that he explored would bear his name. But the politics of naming also had geopolitical significance. In such features as Canada Glacier, Commonwealth Glacier and Wales Glacier, Taylor's party inscribed the region with a British imperial presence.

It was not until several decades after the initial 1923 claim that the New Zealand Government took an active interest in promoting its Antarctic sovereignty. From the late 1950s, the early names given to geographical features in the McMurdo Dry Valleys would prove helpful in making a case for sovereignty. Although the Heroic Era place names were connected to the British Empire in general rather than to New Zealand specifically, imperial affiliations meant that New Zealanders could use these names to extend a sense of their presence in the region further back in time. (This was an age when Edmund Hillary made the first successful ascent of Mount Everest, both as a New Zealander and a proud subject of the British Commonwealth.) Even the name McMurdo Dry Valleys – which was commonly used from the late 1950s – connected the region to early British interests in the continent: the sound adjacent to the valleys had been named after Archibald McMurdo, an officer on the James Clark Ross's British expedition to Antarctica between 1839 and 1843. Recognising the importance of naming places in a region they claimed to own, the New Zealand government wasted little time in giving place names to other parts of the McMurdo Dry Valleys (Bull & Barwick 2009). Victoria Valley, for example, was named after the Victoria University of Wellington (Alberts 1995, p. 783).

In rejecting New Zealand's claim to the Ross Dependency, the US government has walked a fine line between scientific cooperation and political rivalry (Joyner 1997). The practical relationship between New Zealand and the United States in Antarctica has been described as one of partnership, despite the fact that the underlying dispute over ownership remains unresolved (Peat 2007). In the construction of place in the McMurdo Dry Valleys, the naming of major features by the British expeditions of the Heroic Era could be seen as working against US political interests. But the United States has, in fact, proved remarkably successful in appropriating the myths of the Heroic Era into its own political interests in the southern continent, as demonstrated by naming the US station at the South Pole Amundsen-Scott Station (Leane 2016, p. 83). In presenting itself as a successor to the Heroic Era in Antarctica, the United States has perhaps benefited from the fact that it had no major expeditions of its own to Antarctica during this period, meaning that it is not historically tied to any particular location. An example of this is the large US McMurdo Station, which has associations with the Heroic Era through the fact that Captain Scott's Discovery Hut is located nearby. Even the name McMurdo Station might be seen as an appropriation of the Antarctic past by the United States, connecting to the expedition of James Clark Ross and transferring readily to the McMurdo Dry Valleys. In using place names to support their political claims, the Americans did not rely solely on appropriating names from the Heroic Era, but also gave their own names to geographical features in the McMurdo Dry Valleys. Don Juan Pond, for example – which turns out to be the most saline body of water on the planet – is named after two US helicopter pilots called Don and John (Alberts 1995, p. 194).

Japanese researchers began work in the McMurdo Dry Valleys in the early 1960s, not long after New Zealand and the United States had renewed research in the region during the IGY. Unlike their English-speaking counterparts, however, Japanese researchers did not give internationally accepted Japanese names to the geographical features of the McMurdo Dry Valleys. This was partly a result of arriving in the region slightly later than the New Zealanders and Americans. It also reflected the fact that English (and not Japanese) is the international language of Antarctic science. Moreover, the Japanese appeared less keen than other nations to establish a permanent presence on the continent. Instead of creating their own scientific station in the region, Japanese researchers shared facilities and logistics with New Zealand and the United States, which might also have led to a reluctance to propose their own place names.

The political consequences of the history of place-naming in the McMurdo Dry Valleys are impossible to state definitively since they are so interconnected with other factors. But it is probably not a coincidence that New Zealand and the United States, the two countries that engaged most energetically in naming geographical features in the McMurdo Dry Valleys, both retain an active presence there, whereas Japan no longer has any scientific interests in the region. Japanese interests came to an abrupt end in the mid-1980s, and since then have largely been forgotten. In thinking about the political importance of place names, it is not implausible to suggest that working in an environment with foreign language place names might have contributed to a sense of alienation experienced by Japanese scientists. And when combined with other factors – such as the fact that most Japanese Antarctic interests were concentrated in other parts of the continent – this sense of alienation might have contributed to the abrupt cessation of Japanese research in the McMurdo Dry Valleys.

The science and technology of place

Under the terms of the 1959 Antarctic Treaty, the connections between science and politics are more explicit than in almost any other part of the world (Howkins 2017a, pp. 148–151). As a consequence, it is difficult to separate scientific research in Antarctica from the underlying political motivations. But in thinking about the construction of place, science and technology have played distinctive roles. As more has been learned about Antarctica, understandings of place have changed quite radically. Scientific advances in fields like microbiology have led to new ways of understanding the environment: Scott's 'Valley of the Dead', for example, has become a centre of ecological research. NASA's recurrent interest in the McMurdo Dry Valleys since the 1960s suggests that the region could be thought of as an analogue for the Moon or Mars, giving a very modern 'space age' sense of place to a remote region (Doran, Lyons & McKnight 2010).

New technologies have played a particularly important role in constructing a sense of place in the McMurdo Dry Valleys. When Scott and Taylor first entered the region in the early twentieth century, they were on foot and only saw the valley that is known today as Taylor Valley. During the US Operation Highjump expedition immediately after the Second World War, aerial photographs taken by Admiral Byrd on a flight over the Dry Valleys showed that the extent of the ice-free region was much larger than just one valley (Harrowfield & New Zealand Antarctic Society 1999). Proper mapping of the region as a whole began during the IGY of 1957–58, as expeditions from New Zealand and the United States started to build up an accurate understanding of the area. But discrepancies over the extent of the region continued. From the 1980s, satellite technology was routinely applied to the scientific investigation of Antarctica. Remote sensing, using platforms such as Landsat, allowed for a new perspective on the Valleys, making it easier to differentiate between land and ice. More recently, Joe Levy (2012, p. 120) used Landsat Image data to state definitively that, 'The greater MDV have a total area of 22700 km^2 and an ice-free area of 4500 km^2', and this figure seems to have become generally accepted, at least for the moment.

For scientists working in the McMurdo Dry Valleys, these increases in the known area of the region have served in a very subtle way to enhance their authority. When researchers say that they have 'done research in the McMurdo Dry Valleys', they really mean that they have done research in certain parts of the McMurdo Dry Valleys. In reality, most scientific research has taken place in the two most accessible valleys: Taylor Valley and Wright Valley. But findings from these areas could then be extended to other parts of the ice-free region, at least in people's minds. Such a process was not particularly rational: the scientists themselves realise, for example, that the high-altitude valleys further inland were often very different from the Taylor and Wright valleys close to the coast. But very few scientific papers on the McMurdo Dry Valleys fail to mention that this is the largest ice-free region on the Antarctic continent, with size perhaps being seen as a proxy for importance.

A lot more explicit than the scientific authority gained from the internal expansion of the known area of the McMurdo Dry Valleys has been the use of scientific research in this region to investigate the Antarctic continent more broadly. One particularly strong example of this is the Dry Valleys Drilling Project (DVDP), a politically and scientifically expedient collaboration between New Zealand, the United States and Japan in the early to mid-1970s. The DVDP was the first real 'big science' project to take place in the McMurdo Dry Valleys, and much of its rationale was based on the idea that research in this part of Antarctica could provide valuable information about the continent as a whole. Lyle McGinnis, the American project coordinator of the DVDP, summed up the idea of the McMurdo Dry Valleys as a 'type area' through a comparison with his home state of Illinois:

> Although preglacial geology, physiology, and climate of the regions are quite different, they each at one time occupied an interlobate zone where continental ice sheets from the east and west deposited their sediments in interfingered sequences that present unique possibilities for correlation ... Illinois has become a classical 'type' area in which the Pleistocene of North America can be interpreted ... The dry valleys too are rapidly becoming the 'type' area for glacial historical studies of Antarctica ...
>
> (McGinnis n.d.)

The idea in both cases (Illinois and the McMurdo Dry Valleys) was that much broader processes could be investigated by studying a relatively small area.

Drilling work for the DVDP commenced in 1972–73 and continued for four summers, with most of the work concentrated in the 1973–74 and 1974–75 seasons. Drilling often did not go as smoothly as intended, and scientific objectives frequently had to be changed in the field. As George Llano, the acting chief scientist at the National Science Foundation's Office of Polar Programs (NSF OPP), noted:

> The first drill site had been at Lake Vanda where about 30 feet of drill casing were lost and barely any core was retrieved. This was an expensive fiasco. Next, Don Juan Pond, also another fiasco, since Dr Keros Cartwright had predicted high pressure ground water. This came true and drilling had to be stopped: no core. At Lake Fryxell calcium chloride muds caused localized melting of permafrost and again relatively little penetration and core retrieval.
>
> (Llano 1974)

Despite these difficulties, a total of fifteen boreholes, varying in depth from two metres to 328 metres, were drilled. Total drill penetration was 2,231 metres, with 2,074 metres of core recovered (McGinnis n.d.), meaning that there was plenty of material to work with as DVDP scientists started to process their results.

Results of the DVDP were discussed at a series of three workshops held in Seattle (1974), Wellington (1976) and Tokyo (1978) and published in a research series titled *Dry Valley Drilling Project Bulletin* and an American Geophysical Union publication (McGinnis 1981). It is not easy to summarise succinctly the results of the DVDP, and this in itself attests to the difficulties in addressing the central research questions (and perhaps says something about the fact that this was not really hypothesis-driven science). But a tremendous amount of data was produced, and some interesting things were learned. The crystalline basement rock in the Dry Valleys, for example, was found to go back to the Cambrian period (541–485 million years ago). In the results of the DVDP, interesting observations were made about the advance and retreat of the West Antarctic and East Antarctic Ice Sheets, although there was little that was conclusive about ice-sheet stability (Sugden 1996). Alongside the logistical lessons learned by the DVDP, one of its legacies might be seen as the effectiveness of the strategy of making broad scientific claims from research in a particular place in terms of winning funding.

The management of place

One interesting consequence of the Dry Valleys Drilling Project was that it spurred new ways of thinking about environmental impact and, by extension, discussions over the best ways to manage the region. By the time that the DVDP was getting started in the early 1970s, biology and ecology were becoming increasingly established fields in the McMurdo Dry Valleys. Although ecology featured in some of the original DVDP research questions, ecologists working in the McMurdo Dry Valleys generally took a negative view of the enterprise as a result of its potential for disturbing the environment. The DVDP coincided with an upsurge of environmental consciousness, both globally and in Antarctica (McCormick 1989). Globally, the early 1970s was the period of the first Earth Day, the United Nations Conference on the Human Environment and, importantly for the history of the Dry Valleys, the early implementation of the National Environment Policy Act in the United States, which had been signed in 1969. In Antarctica, after a decade of science conducted within the framework of the Antarctic Treaty, there was a growing awareness of the impact of scientific research on the Antarctic environment.

One of the first discussions of the DVDP among ecologists took place at a Colloquium on Conservation Problems in Antarctica held at Virginia Polytechnic Institute and State University (now commonly known as Virginia Tech). The proceedings of the colloquium noted:

> As informally discussed by several participants of this colloquium, the need for careful advanced planning is exemplified by the proposed DVPD. Apparently, geologists are eager to undertake this project without much

understanding of the possible impact which the DVDP will have on the delicate, unique, and scientifically indispensable soil and freshwater eco-systems of the dry valleys.

(Llano 1974)

Among the most concerned biologists were Roy Cameron and Bruce Parker, who were both active in the Dry Valleys, along with George Llano, the acting chief scientist at the NSF OPP. Llano sought wider opinions and noted that, 'All those replying generally reinforced our apprehensions about DVDP's environmental impact. Several went further to elaborate that DVDP was not good science, a waste of OPP/NSF funds and even questioning the basis for selection of DVDP presence' (Llano 1974).

Although pressure from higher up in the Office of Polar Programs – likely influenced by the diplomatic importance of the DVDP – meant that drilling went ahead, concerns about the environment did not go away. George Llano in parti-cular continued to push for ecological oversight of the drilling project. In April 1973, Roy Cameron and Bruce Parker were invited to Dekalb, Illinois, to co-chair a committee to formulate an Environmental Impact Statement (EIS) for the DVDP. Although this work was a little late for the 1973–74 season, pressure from biologists and others was able to prevent Lake Bonney – one of the major sites for early limnological investigation – from being added to the DVDP program for the following year. And by the 1974–75 drill season, the EIS system was working more effectively, with biologists from Virginia Polytechnic Institute and State University having full responsibility for monitoring with support from the drillers (Llano 1974). At the end of the DVDP, Lyle McGinnis himself – despite initial reluctance and even hostility towards environmental monitoring – was listing the environmental efforts as one of the success stories of the whole enterprise.

The use of EISs and subsequent monitoring for regulating the DVDP not only provided a model for the future environmental management of the McMurdo Dry Valleys, but also became part of the environmental manage-ment strategies for the Antarctic continent as a whole. Under the terms of the Madrid Protocol, signed in 1991 and ratified in 1998, all activities in Antarctica are required to go through an environmental impact assessment. The Madrid Protocol also provides for the designation of Antarctic Specially Managed Areas (ASMAs) and Antarctic Specially Protected Areas (ASPAs). In a neat circularity, in 2004 the McMurdo Dry Valleys were adopted as ASMA number 2 within the Antarctic Treaty System, following a joint proposal from the United States and New Zealand. The creation of the McMurdo Dry Valleys ASMA high-lights the connections between the local, continental and global in the con-struction of a sense of place in Antarctica. Thirty years after unease about the environmental impact of the DVDP had led to the use of EISs, a continent-wide policy came back to manage this area, connecting the McMurdo Dry Valleys to the authority of the Antarctic Treaty System and to what anthro-pologist Jessica O'Reilly (2017) has labelled the 'technocratic Antarctic'.

Figure 11.1 Map of McMurdo Dry Valleys Antarctic Specially Managed Area (ASMA No. 2)
Source: Polar Geospatial Center, 2018, '*PGC Map Catalog*', <https://doi.org/10.7910/DVN/6R8F7U>, Harvard Dataverse, V1, ANT REF-ES2004–001, accessed 6 February 2019

As part of the management strategy for the McMurdo Dry Valleys, the countries responsible for the ASMA have produced a map to show the boundaries of the protected area. This map is periodically revised to show updates to the management regime. In many ways, the ASMA map is one of the best representations of place in the contemporary McMurdo Dry Valleys. ASMAs and ASPAs have political implications, and politics, science, technology and the environment are intimately connected within them. In a continent where the scope for political action is somewhat constrained, environmental management offers one effective way of demonstrating an active political interest, and the perceived success of environmental protection measures within any given region takes on a heightened significance (Howkins 2017b). More broadly, the success of the Madrid Protocol as a whole is determined by the functioning of environmental protection in multiple distinct places.

Conclusions

Even a brief overview of some of the ways in which place has been constructed in the McMurdo Dry Valleys shows that it is inextricably connected to the wider history of the region. In fact, it is almost impossible to make any statement about place in the McMurdo Dry Valleys without taking some sort of position on its political history and the associated histories of science, technology and environmental management. Just to mention Taylor Valley or give a precise spatial extent of the ice-free area is to take a position in relation to the history. Such an observation means that we frequently risk anachronism and teleology in defining the places we study. It is not quite correct, for example, to say that Captain Scott discovered the McMurdo Dry Valleys: he was simply the leader of the first expedition to sight one part of the region that would later become known as the McMurdo Dry Valleys. But what we might lose in terms of definitional consistency, we gain through the insight that the construction of place is embedded into the wider histories we study.

An answer to the question of how we can write engaging, relevant histories of Antarctica in places such as the McMurdo Dry Valleys without falling into the trap of flattening the rest of the continent might focus on embracing both distinctiveness and connectivity. Place in the McMurdo Dry Valleys has been created through the interaction of different forces acting at local, continental and global scales. This works both ways. A focus on the uniqueness of the region should not preclude a degree of extrapolating outwards – in fact, such extrapolation has very much been part of the construction of the McMurdo Dry Valleys as a place. At the same time, the region has been shaped by wider trends, such as the emerging global environmentalism of the 1970s and 1980s. There remains a fundamental specificity to the history of place in the region, created by the unique (and fluid) overlaps with other parts of the Antarctic continent and broader global trends.

What does this tell us about the idea of the Anthropocene in the history of Antarctica? As noted in the introduction to this chapter, the concept of the Anthropocene has been surprisingly absent from the history of the McMurdo Dry

Valleys. For all the ways that researchers have used the McMurdo Dry Valleys to make broader claims for the relevance of their work, even today the Anthropocene is a term that is seldom used by scientists working in the region. While this certainly reflects the novelty of the term, it also reflects the difficulty of applying such a broad-brush concept to study of a unique region, where coming to terms with local specificity is one of the key challenges of scientific research. Ultimately, the problem of geographical specificity in Antarctic history is perhaps best viewed as an opportunity. As we start to take place in Antarctica more seriously, it is important to highlight local specificity. But as the history of the McMurdo Dry Valleys demonstrates, local specificity does not mean we need to study places as disconnected from wider trends. In fact, these wider trends have played an important role in the construction of local place. This certainly leaves room for the concept of the Anthropocene, but also poses a challenge to it. Anthropogenic global change can only be one factor among many that contribute to the construction of a place like the McMurdo Dry Valleys, and any attempt to over-emphasise its importance is likely to have political implications.

References

Alberts, FG 1995, *Geographic place names in Antarctica*, 2nd edn, National Science Foundation, Arlington.

Antonello, A 2016, 'Finding place in Antarctica' in P Roberts, L-M van der Watt & A Howkins (eds), *Antarctica and the humanities*, Palgrave Studies in the History of Science and Technology, Palgrave Macmillan, London, pp. 181–204.

Bocking, S 2007, 'Science and spaces in the northern environment', *Environmental History*, vol. 12, pp. 867–894.

Bowler, PJ 1993, *The Norton history of the environmental sciences*, Norton History of Science, WW Norton, New York.

British Antarctic Survey 2015, *British Antarctic Survey*, viewed 21 December 2018, <https://www.bas.ac.uk/>.

Bull, C & Barwick, D 2009, *Innocents in the Dry Valleys: an account of the Victoria University of Wellington Antarctic expedition, 1958–59*, Victoria University Press, Wellington.

Carey, M, Garone, P, Howkins, A, Endfield, G, Culver, L, Johnson, S, White, S & Fleming, JR 2014, 'Forum: Climate change and environmental history', *Environmental History*, pp. 281–364.

Debenham, F 1923, *British (Terra Nova) Antarctic Expedition 1910–1913: report on the maps and surveys*, Harrison & Sons, London.

Dodds, K 2002, *Pink ice: Britain and the South Atlantic empire*, IB Tauris, London.

Doran, PT, Berry Lyons, W & McKnight, DM 2010, *Life in Antarctic deserts and other cold dry environments*, Astrobiological Analogs, Cambridge University Press, Cambridge & New York.

Fontana, P 2014, *La pugna Antártica, el conflicto por el sexto continente: 1939–1959*, Guazuvira Ediciones, Buenos Aires.

Harrowfield, DL & New Zealand Antarctic Society 1999, *Vanda Station: history of an Antarctic outpost, 1968–1995*, New Zealand Antarctic Society, Christchurch.

Howkins, A 2016, 'Taylor's Valley: what the history of Antarctica's "Heroic Era" can contribute to contemporary ecological research in the McMurdo Dry Valleys', *Environment and History*, vol. 22, pp. 3–28, doi:10.3197/096734016x14497391602125.

Howkins, A 2017a, *Frozen empires: an environmental history of the Antarctic Peninsula*, Oxford University Press, New York.

Howkins, A 2017b, 'Politics and environmental regulation in Antarctica: a historical perspective' in K Dodds, AD Hemmings & P Roberts (eds), *Handbook on the politics of Antarctica*, Edward Elgar, Cheltenham, pp. 337–350.

Joyner, CC 1997, *Eagle over the ice: the US in the Antarctic*, University Press of New England, Hanover.

Leane, E 2016, *South Pole: Nature and Culture*, Reaktion Books, London.

Lecain, T 2015, 'Against the Anthropocene. A neo-materialist perspective', *International Journal for History, Culture and Modernity*, vol. 3, pp. 1–28, doi:10.18352/hcm.474.

Levy, J 2012, 'How big are the McMurdo Dry Valleys? Estimating ice-free area using Landsat Image data', *Antarctic Science*, vol. 25, pp. 119–120, doi:10.1017/s0954102012000727.

Llano, G 1974, 'Office of Polar Programs Diary Note, 11 November 1974', Mort Turner Papers, box 36, folder 6, Byrd Polar and Climate Research Center Archival Program, Ohio State University, Columbus.

McCormick, J 1989, *Reclaiming paradise: The global environmental movement*, Indiana University Press, Bloomington.

McGinnis, LD n.d., 'Review of the DVDP (1971–1977)', Mort Turner Papers, box 37, folder 14, Byrd Polar and Climate Research Center Archival Program, Ohio State University, Columbus.

McGinnis, LD (ed.) 1981, *Dry Valley Drilling Project*, American Geophysical Union, Washington, D.C.

O'Reilly, J 2017, *The technocratic Antarctic: An ethnography of scientific expertise and environmental governance*, Cornell University Press, Ithaca.

Peat, N 2007, *Antarctic partners: 50 years of New Zealand and United States cooperation in Antarctica, 1957–2007*, Phantom House, Wellington.

Priscu, JC 1998, *Ecosystem dynamics in a polar desert: the McMurdo Dry Valleys, Antarctica*, Antarctic Research Series, vol. 72, American Geophysical Union, Washington, D.C.

Scott, RF 2001, *The voyage of the 'Discovery'*, 2 vols, Cooper Square Press, New York.

Sugden, DE 1996, 'The East Antarctic Ice Sheet: unstable ice or unstable ideas?', *Transactions of the Institute of British Geographers*, vol. 21, no. 3, pp. 443–454.

Templeton, M 2000, *A wise adventure: New Zealand in Antarctica, 1920–1960*, Victoria University Press, Wellington.

Part 4
Conclusion

12 Antarctica looking forward

Four themes

Jeffrey McGee and Elizabeth Leane

The 'age of humans' is significantly changing Antarctica and the Southern Ocean. Through different issues and disciplinary lenses, the contributions to this volume have explored the causes, effects and meanings of human interaction with the region. In this concluding chapter, we draw out some insights from these contributions as a group, suggesting four interrelated themes that capture the various ways in which the relationship between humans and the Antarctic continent is currently being perceived.

The first of these themes is 'Antarctica as Near'. We use this phrase to foreground the ongoing erosion of barriers of physical separation between human activity and Antarctica. Traditionally conceived as the 'end of the Earth', more remote even than the high Arctic, Antarctica has been considered isolated and separate from the rest of the human world. Until the establishment of permanent scientific research bases in the mid-twentieth century, there was only limited human contact with the Antarctic continent. Before this time, most human contact was periodic and related to sealing and whaling activities in the Southern Ocean and various short-term discovery and mapping expeditions. As discussed in the introduction, there are now nearly eighty research stations located on the Antarctic continent and upwards of fifty thousand tourists visiting the continent each year. Direct air flights supplying research bases in Antarctica, as well as tourist flights to and over the continent, are regular features of contemporary Antarctic travel and habitation. The icefish, toothfish and krill fisheries of the Southern Ocean are governed by a treaty (the CAMLR Convention) that is a leading light in the field of international environmental governance, although there has recently been tension over the formation of Marine Protected Areas. The traditional notion of separation between humanity and the Antarctic region is therefore under significant strain on many fronts.

The opening chapter by Tim Stephens focuses upon the extended causal links between human activity in emitting greenhouse gases and resultant biophysical change in the Antarctic continent. It argues that the Anthropocene highlights the need for a better fit between governance systems and these global drivers of biophysical change in the region. Carolyn Philpott points to the way in which contemporary soundscape composers have begun to incorporate anthropogenic as well as environmental sound into their artistic works, as a way of exploring the increasing entanglement of the human and nonhuman in the Antarctic region. Other contributors question conventional ways of thinking about Antarctic human history and particularly the stereotypical image of the only continent without a historical population. For instance, the chapters by Ben Maddison and Nelson Llanos respectively uncover participation of indigenous peoples in Antarctic exploration and early commerce, and more recent Chilean efforts at permanent human settlement.

The second theme is 'Antarctica as Vulnerable'. This theme speaks directly to the large-scale biophysical threats to Antarctica and the Southern Ocean from human activities. In the recent past, perhaps as late as the 1950s, the environmental impacts of human activities on Antarctica and the surrounding Southern Ocean were predominantly observed through sealing, whaling, the construction of scientific research stations and exploratory activities. However, global environmental problems such as thinning of the ozone layer, human-induced climate change and ocean acidification are now significant drivers of biophysical change in the region. These manifest as threats to the Antarctic continent through the loss of ice sheets and damaged marine ecosystems. The largest and most serious longer-term threats to the Antarctic environment are now generated by human activities in population centres far removed from the relative geographic isolation of the Antarctic continent and Southern Ocean. The foreword by Sanjay Chaturvedi articulates the intersections between economics, politics and law at a global scale that are driving forces of significant biophysical change in the Antarctic continent and Southern Ocean. Tim Stephens' chapter highlights the intersections between the Antarctic Treaty System and wider systems of global governance that might be called upon to respond to these threats. The chapter by Jeffrey McGee illustrates how human discourses of geoengineering are generating ideas to protect the Antarctic continent from the threat of climate change through ocean sequestration of carbon, marine cloud brightening and the use of reflective particles. McGee's chapter also highlights an Antarctic glacial stabilisation discourse that proposes mega-scale engineering projects to help stabilise Antarctic ice sheets from deterioration and eventual loss. Given the scale of these glacial stablisation proposals, they may in themselves represent significant biophysical risks to the Antarctic continent. In the popular cultural sphere, Hanne Nielsen critically analyses the way in which the image of Antarctica as a threatened environment is commodified for commercial purposes.

The third theme identified is 'Antarctica as Rescuer'. Under the governance of an international treaty that prioritises scientific research and exchange, the Antarctic is closely identified with the generation of new knowledge to help understand and solve the global environmental problems mentioned above, such as climate change, ozone depletion and ocean acidification. As much as these global environment problems are key drivers of change, scientific work in Antarctica and the Southern Ocean is also seen as central to producing the knowledge needed for humans to respond effectively to them. The Antarctic Treaty System is built on values of peaceful use, scientific investigation and environmental protection of the Antarctic continent. It has long been viewed as one of the more successful examples of international environmental governance and treaty law. This perhaps utopian view of the Antarctic Treaty System sees the approach to managing the continent as a potential model for other international disputes. This theme is critically explored in the chapters by Alan Hemmings and Adrian Howkins, particularly in relation to programmes for gathering scientific knowledge through drilling into the Antarctic ice sheet and terrestrial areas of the McMurdo Dry Valleys. These chapters illustrate the expectations for Antarctica as a global scientific laboratory and also the connections of Antarctic science with geopolitical concerns regarding territorial claims.

The fourth and final theme is 'Antarctica as Threat'. This theme highlights the potential dangers which the Antarctic continent represents to the rest of the planet as a source of future sea-level rise and as a location of extreme weather and climatic conditions that can affect humanity. Leane's chapter examines the way this sense of threat operates in thriller narratives, in which the Antarctic icescape becomes an actor in human dramas, entangled with villains and heroes, and is even itself weaponised. The chapters by Juan Francisco Salazar, Jeffrey McGee and Tim Stephens raise the geopolitical and governance challenges of a melting Antarctic continent under conditions of climate change, which might have profound impacts upon countries with little or no contact with Antarctic concerns or issues. This threatening representation of Antarctic futures points to ominous aspects of the Antarctic ice sheets and their potential impacts on the coastlines of the rest of the planet.

We hope that this collection will serve as an interdisciplinary foundation for further thinking and analysis of the interconnections between humans and the Antarctic continent and Southern Ocean. We believe the collection shows that the concept of the Anthropocene, contested as it is, can nonetheless serve an important role in Antarctic research. Its theoretical and political limitations notwithstanding, the Anthropocene provides a framework for teasing out the various ways in which the fate of humanity and Antarctica are being drawn closer together and for addressing the challenges this poses to our current understandings of the continent and systems of governance of the region.

Jeffrey McGee and Elizabeth Leane
Hobart, Tasmania
April 2019

Index